my **revision** notes

OCR A2
BIOLOGY

Frank Sochacki

HODDER
EDUCATION

Hodder Education, an Hachette UK company, 338 Euston Road, London NW1 3BH

Orders

Bookpoint Ltd, 130 Milton Park, Abingdon, Oxfordshire OX14 4SB

tel: 01235 827827

fax: 01235 400401

e-mail: education@bookpoint.co.uk

Lines are open 9.00 a.m.–5.00 p.m., Monday to Saturday, with a 24-hour message answering service. You can also order through the Hodder Education website: www.hoddereducation.co.uk

ISBN 978-1-4441-7973-6

First printed 2013
Impression number 5 4 3 2 1
Year 2018 2017 2016 2015 2014 2013

Cover photo reproduced by permission of peter_waters/Fotolia
Other photos © Richard Fosbery

Typeset by Datapage (India) Pvt. Ltd.
Printed in India

Hachette UK's policy is to use papers that are natural, renewable and recyclable products and made from wood grown in sustainable forests. The logging and manufacturing processes are expected to conform to the environmental regulations of the country of origin.

P2187

Get the most from this book

Everyone has to decide his or her own revision strategy, but it is essential to review your work, learn it and test your understanding. These Revision Notes will help you to do that in a planned way, topic by topic. Use this book as the cornerstone of your revision and don't hesitate to write in it — personalise your notes and check your progress by ticking off each section as you revise.

☑ **Tick to track your progress**

Use the revision planner on pages 4 and 5 to plan your revision, topic by topic. Tick each box when you have:

● revised and understood a topic
● tested yourself
● practised the exam questions and gone online to check your answers and complete the quick quizzes

You can also keep track of your revision by ticking off each topic heading in the book. You may find it helpful to add your own notes as you work through each topic.

My revision planner

Unit F214 Communication, homeostasis and energy

	Revised	Tested	Exam ready
1 Communication			
7 Monitoring and responding to changes	☐	☐	☐
2 Nerves			
11 Converting energy from a stimulus	☐	☐	☐
15 Transmission at synapses	☐	☐	☐

Meiosis

Meiosis ———————————————————————— Revised ☐

Meiosis is the division of the nucleus to produce four haploid nuclei. Before division starts (during interphase), the DNA replicates so that each **chromosome** consists of two identical copies called sister chromatids that are held together by a centromere. The cell divides twice to produce four daughter cells, each of which contains half the number of

Examiner's tip
This examination paper includes synoptic marks. These test your:
● understanding of the principles behind different processes
● ability to make links back to

Features to help you succeed

Examiner's tips and summaries

Expert tips are given throughout the book to help you polish your exam technique in order to maximise your chances in the exam. The summaries provide a quick-check bullet list for each topic.

Typical mistakes

The author identifies the typical mistakes candidates make and explains how you can avoid them.

Revision activities

These activities will help you to understand each topic in an interactive way.

Now test yourself

These short, knowledge-based questions provide the first step in testing your learning. Answers are at the back of the book.

Exam practice

Practice exam questions are provided for each topic. Use them to consolidate your revision and practise your exam skills.

Definitions and key words

Clear, concise definitions of essential key terms are provided on the page where they appear. Key words from the specification are highlighted in bold for you throughout the book.

Online

Go online to check your answers to the exam questions and try out the extra quick quizzes at **www.therevisionbutton.co.uk/myrevisionnotes**

My revision planner

_J70817_
574 Soc
CNC.

Unit F214 Communication, homeostasis and energy

	Revised	Tested	Exam ready
1 Communication			
7 Monitoring and responding to changes	☐	☐	☐
2 Nerves			
11 Converting energy from a stimulus	☐	☐	☐
15 Transmission at synapses	☐	☐	☐
3 Hormones			
18 Hormone action	☐	☐	☐
21 Regulating blood glucose and insulin	☐	☐	☐
4 Excretion			
25 Removing metabolic wastes	☐	☐	☐
32 Control of water potential	☐	☐	☐
33 Renal failure and urine samples	☐	☐	☐
5 Photosynthesis			
36 The mechanisms of photosynthesis	☐	☐	☐
38 Stage 1: Converting light energy to chemical energy	☐	☐	☐
40 Stage 2: The Calvin cycle	☐	☐	☐
41 Factors affecting the rate of photosynthesis	☐	☐	☐
6 Respiration			
45 Transfer of energy to ATP	☐	☐	☐
46 Aerobic respiration	☐	☐	☐
51 Anaerobic respiration	☐	☐	☐

Unit F215 Control, genomes and environment

	Revised	Tested	Exam ready
7 Cellular control			
53 Coding for proteins	☐	☐	☐
58 Genes that control development	☐	☐	☐
8 Meiosis and variation			
60 Meiosis	☐	☐	☐
63 Genetic diagrams	☐	☐	☐
65 Interactions between loci	☐	☐	☐
67 Variation	☐	☐	☐
69 Selection and genetic drift	☐	☐	☐

Exam practice answers and quick quizzes at **www.therevisionbutton.co.uk/myrevisionnotes**

	Revised	Tested	Exam ready
9 Cloning in plants and animals			
74 Types of cloning	☐	☐	☐
74 Cloning plants	☐	☐	☐
76 Cloning animals	☐	☐	☐
10 Biotechnology			
79 Using microorganisms	☐	☐	☐
80 Using enzymes	☐	☐	☐
81 Growing conditions	☐	☐	☐
11 Genomes and gene technologies			
84 Genome sequencing	☐	☐	☐
85 Genetic manipulation	☐	☐	☐
12 Ecosystems			
92 Ecosystems as complex interactions	☐	☐	☐
93 Energy transfer through ecosystems	☐	☐	☐
95 Ecosystems as dynamic entities	☐	☐	☐
13 Populations and sustainability			
98 Population size	☐	☐	☐
99 Managing ecosystems	☐	☐	☐
14 Plant responses			
103 Responding to environmental changes	☐	☐	☐
15 Animal responses			
108 Responding to environmental changes	☐	☐	☐
111 Movement and muscular coordination	☐	☐	☐
16 Animal behaviour			
118 Innate and learned behaviours	☐	☐	☐
120 Understanding human behaviour	☐	☐	☐

122 Now test yourself answers

Exam practice answers and quick quizzes at www.therevisionbutton.co.uk/myrevisionnotes

Countdown to my exams

6–8 weeks to go

- Start by looking at the specification — make sure you know exactly what material you need to revise and the style of the examination. Use the revision planner on pages 4 and 5 to familiarise yourself with the topics.
- Organise your notes, making sure you have covered everything on the specification. The revision planner will help you to group your notes into topics.
- Work out a realistic revision plan that will allow you time for relaxation. Set aside days and times for all the subjects that you need to study, and stick to your timetable.
- Set yourself sensible targets. Break your revision down into focused sessions of around 40 minutes, divided by breaks. These Revision Notes organise the basic facts into short, memorable sections to make revising easier.

Revised ☐

4–6 weeks to go

- Read through the relevant sections of this book and refer to the examiner's tips, examiner's summaries, typical mistakes and key terms. Tick off the topics as you feel confident about them. Highlight those topics you find difficult and look at them again in detail.
- Test your understanding of each topic by working through the 'Now test yourself' questions in the book. Look up the answers at the back of the book.
- Make a note of any problem areas as you revise, and ask your teacher to go over these in class.
- Look at past papers. They are one of the best ways to revise and practise your exam skills. Write or prepare planned answers to the exam practice questions provided in this book. Check your answers online and try out the extra quick quizzes at **www.therevisionbutton.co.uk/myrevisionnotes**
- Use the revision activities to try different revision methods. For example, you can make notes using mind maps, spider diagrams or flash cards.
- Track your progress using the revision planner and give yourself a reward when you have achieved your target.

Revised ☐

One week to go

- Try to fit in at least one more timed practice of an entire past paper and seek feedback from your teacher, comparing your work closely with the mark scheme.
- Check the revision planner to make sure you haven't missed out any topics. Brush up on any areas of difficulty by talking them over with a friend or getting help from your teacher.
- Attend any revision classes put on by your teacher. Remember, he or she is an expert at preparing people for examinations.

Revised ☐

The day before the examination

- Flick through these Revision Notes for useful reminders, for example the examiner's tips, examiner's summaries, typical mistakes and key terms.
- Check the time and place of your examination.
- Make sure you have everything you need — extra pens and pencils, tissues, a watch, bottled water, sweets.
- Allow some time to relax and have an early night to ensure you are fresh and alert for the examination.

Revised ☐

My exams

A2 Biology Unit F214

Date:..

Time: ....

Location:...

A2 Biology Unit F215

Date: ....

Time: ....

Location:...

1 Communication

Monitoring and responding to changes

The need for communication between cells

Revised

In **multicellular organisms**, the cells specialise to perform certain tasks more efficiently. These tasks include gaseous exchange, absorbing nutrients, removing wastes, etc. Any one individual cell from a multicellular organism may not be able to perform all the tasks required for survival — the cells rely on one another.

The organism must maintain a relatively narrow range of internal conditions in which enzymes can work and enable cell processes to continue. Therefore, receptor cells are required that detect changes in both the **internal conditions** and the **external environment**. If changes occur that will alter the ability of enzymes to function, these changes act as stimuli and the receptor cells must communicate with other cells in the organism to respond appropriately. This response may be to move away from harmful external conditions or to reverse an internal process that is bringing about the change. Such responses may require the **coordination** of several organs — for example, running away from a predator involves the nervous system, the skeleton and muscles, the lungs and the circulatory system as well as increases in respiration within the cells.

Revision activity

Write a list of all the external conditions that may act as stimuli to bring about a response.

Cell signalling

Revised

Cells communicate with each other by a process called **cell signalling**. The **nervous system** and the **endocrine (hormonal) system** are both examples of cell signalling. The nervous system is a dedicated communication system that spreads throughout the body, whereas the hormonal system uses the circulatory system to transport signalling molecules called hormones. Local communication can be achieved by hormone-like molecules that diffuse through the tissue fluid, such as cytokines and histamine in the immune response.

Cell signalling is communication between cells.

Feedback and homeostasis

Negative feedback

Systems in the body monitor certain parameters such as internal temperature. These parameters need to be kept at a set point that corresponds to the optimum conditions. If the internal temperature changes away from the set point, a mechanism is put in place to reverse the change. This ensures that the internal temperature does not change too much and remains fairly constant. This is **negative feedback**.

> **Negative feedback** is the response to a change in conditions that acts to reverse that change.

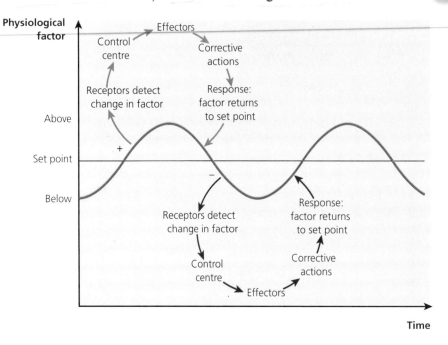

Figure 1.1 The principle of negative feedback

Positive feedback

If a parameter such as temperature changes, it may bring about further increases in that change. For example, as bacteria respire they produce heat and the increased temperature makes the bacteria more active so they respire more and release even more heat. **Positive feedback** is usually harmful, but there are one or two examples where it can be beneficial. In nervous conduction a small change in membrane potential opens ion channels that allow the movement of ions through the membrane, which further increases the change in membrane potential to produce an action potential.

> **Positive feedback** is when a small change brings about an increase in that change.

Homeostasis

Homeostasis is the maintenance of a constant internal environment. The list of parameters that are monitored and maintained includes temperature, water potential of blood, pH, blood glucose concentration, blood pressure, and blood concentration of ions such as Na^+, K^+ and Ca^{2+}. Homeostasis relies on monitoring the levels of these parameters using receptors. A negative feedback mechanism then controls the action of effectors which can reverse any internal change that may occur and return the parameter to its set point.

Now test yourself

1. Explain why positive feedback can only be used for a short period of time in living systems.

2. For each internal parameter that is monitored, write a note explaining why it is must be kept constant. This explanation should include a detailed statement about what will happen if the parameter gets too low or too high.

Answer on p. 122

Tested

Maintaining body temperature

↑ external sources of heat

Ectotherms are organisms that rely on their surroundings to gain body heat. Their body temperature is always dependent on the surrounding environmental temperature. However, many ectotherms are able to regulate their body temperature to some extent, usually by **behavioural** adaptations:

● Lizards may bask in the sun and even flatten their rib cage to gain a larger surface area in order to warm up. They hide in the shade or burrow when too hot.

● Locusts orientate themselves side-on to the sun in order to warm up when they are cold, but climb a plant stem to get away from warm ground when they are too hot.

● Many insects flap their wings to generate some heat in their wing muscles before flying.

Endotherms are organisms that use internal mechanisms to regulate their body temperature. This is known as thermoregulation. Receptors detect changes in the temperature of the environment (**peripheral temperature receptors** in the skin) and in the internal core temperature (the core temperature receptor in the **hypothalamus** in the brain). Both sets of information are received in the thermoregulatory centre in the hypothalamus, which then sends out instructions to make adjustments to reverse any change in core body temperature. The **effectors** that bring about these changes include the circulatory system, the skin, the liver and the muscles (Table 1.1).

> **Ectotherms** are organisms that rely on external sources of heat.

> **Revision activity**
>
> Draw a sketch outline of a reptile and add arrows to indicate each source of heat gain and loss. Annotate the arrows to explain how the heat is gained or lost.

> **Typical mistake**
>
> Many candidates refer to ectotherms as 'cold-blooded' — this is not the case as some can keep their body temperature well above 30°C.

> **Endotherms** generate their own heat to regulate their body temperature more effectively.

Table 1.1 Effectors for adjusting body temperature

Effector	Response if too hot	Response if too cold
Sweat glands	More sweat is produced — evaporation cools the skin	Less sweat is produced
Erector pili muscles in skin (attached to hairs)	Muscles relax, causing the hairs to lie flat and allowing the air to circulate over the skin	Muscles contract, raising the hairs and trapping an insulating layer of air next to the skin
Blood vessels in skin	Vasodilation — blood flows close to skin surface to lose heat by radiation and convection	Vasoconstriction — blood is diverted to flow further from the surface, so radiation and convection are reduced
Muscles	Muscles relax	Muscles contract and relax repeatedly, generating heat by friction and from metabolic reactions — this is shivering
Liver	Reduces metabolic rate	Increases metabolic rate to release heat

> **Typical mistake**
>
> Candidates often refer to blood vessels in the skin moving closer to the surface or deeper down. The blood vessels themselves do not move — they simply dilate or constrict to divert the blood flow.

> **Revision activity**
>
> Draw a sketch outline of a mammal and add arrows to indicate each source of heat gain and loss. Annotate the arrows to explain how the heat is gained or lost.

Now test yourself

Tested

3 Explain why the skin is the main organ of thermoregulation.

Answer on p. 122

Exam practice

1 **(a)** Explain the term *cell signalling*. [2]

(b) Name two systems in the mammalian body that use cell signalling. [2]

(c) Define the term *homeostasis*. [2]

(d) Explain why negative feedback is essential in a homeostatic mechanism. [4]

2 **(a)** In the early morning, bees can be seen buzzing their wings but not flying. Explain the reasons behind this behaviour. [3]

(b) Explain how blood can be diverted to and from the skin to help thermoregulation in a mammal. In your answer you should use appropriate technical terms, spelt correctly. [5]

(c) The graph below shows the rate of flow of heat through the skin of a person during exercise.

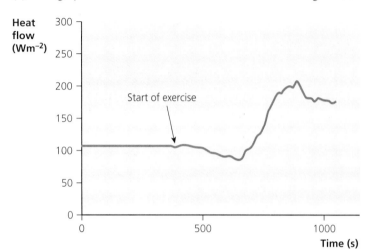

(i) Describe the changes shown in the graph. [3]

(ii) Explain the changes shown in the graph. [4]

Answers and quick quiz 1 online

Online

Examiner's summary

By the end of this chapter you should be able to:

✔ Outline the need for communication systems within multicellular organisms.

✔ State that cells need to communicate with each other by a process called cell signalling.

✔ State that neuronal and endocrine systems are examples of cell signalling.

✔ Define the terms *negative feedback, positive feedback* and *homeostasis*.

✔ Explain the principles of homeostasis.

✔ Describe the responses that maintain core body temperature in ectotherms and endotherms.

2 Nerves

Converting energy from a stimulus

Sensory receptors   Revised ☐

Sensory receptors are essential in monitoring conditions in the environment or within the body. They detect changes in the conditions and only respond when the conditions change. Such a change in conditions is called a **stimulus**. Receptors are transducers, which means they convert the energy from a stimulus into electrical energy in the form of an **action potential**.

There are a wide range of sensory receptors. Each type of receptor can detect a specific energy change. For example, the temperature receptors detect changes in temperature and the Pacinian corpuscles detect changing pressure on the skin caused by movement.

> **Sensory receptors** are specialised cells that convert one form of energy into an action potential.
>
> An **action potential** is the change in membrane potential that is transmitted along the neurone.

Sensory and motor neurones   Revised ☐

Neurones are nerve cells. The two most important types of nerve cell are **sensory neurones** and **motor neurones** (Table 2.1).

> **Neurones** are specialised cells that can transmit an action potential.

Table 2.1 Sensory and motor neurones

Feature	Sensory neurone	Motor neurone
Function	Transmit action potentials from the sensory receptor to the central nervous system	Transmit action potentials from the central nervous system to the effectors such as muscles
Myelin sheath	Both types of neurones possess a myelin sheath to speed up conduction and insulate the neurone from the surrounding cells	
Cell body	Positioned just outside the central nervous system in a swelling called the dorsal root ganglion	Positioned inside the central nervous system
Dendrites	Many	Many
Dendron	One long dendron	No dendron
Axon	One short axon	One long axon
Ending	Synaptic knob	Motor end plate
Diagram		

Revision activity

Sketch diagrams of a sensory and a motor neurone and annotate the diagrams to show how these cells are adapted to their function.

Now test yourself

1 Explain why a motor neurone has a long axon whereas a sensory neurone has a short axon.

Answer on p. 122

Tested

Resting potential

Revised

When a neurone is at rest, it maintains a potential difference in charge across the cell membrane — the membrane is polarised. This is called the **resting potential**. The inside of the cell is maintained at −60 to −70 mV compared to outside the cell. The neuronal membrane contains sodium–potassium ion pumps, which maintain the potential difference. Using ATP, they pump three sodium ions out of the cell for every two potassium ions into the cell. Therefore, while at rest the cell contains many potassium ions and few sodium ions.

> **Resting potential** is the potential difference in charge across the cell membrane when the cell is at rest.

Action potential

When a neurone is stimulated, it becomes depolarised (it loses its normal polarisation). The neuronal membrane contains special gated channels that open as a result of a change in the potential difference — they are **voltage-gated sodium ion channels**. These allow sodium ions to flow by facilitated diffusion into the cell.

As a result, the inside of the cell becomes less negative compared to the outside. If enough sodium ions flow into the cell, the potential across the membrane reaches the threshold potential, which opens more sodium channels — this is positive feedback. More sodium ions flow into the neurone and the potential inside the cell rises to about +40 mV compared to outside. This is known as an action potential.

After the sodium ions have entered the cell, **voltage-gated potassium ion channels** open to allow potassium ions out of the cell. This reduces the potential difference across the membrane again, returning it to the −60 mV resting potential. This is called repolarisation. The membrane potential then briefly falls below the normal resting potential, which is called hyperpolarisation.

Now test yourself

2 Explain why the first part of an action potential is called depolarisation and the second part is repolarisation.

Answer on p. 122

Tested

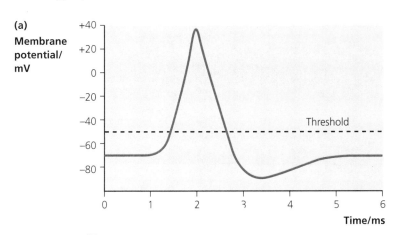

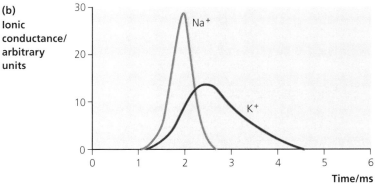

Figure 2.1 The change in (a) potential difference and (b) the conductance of sodium and potassium ions during the passage of an action potential

Typical mistake

Many candidates seem unsure how the ions move across the membrane during an action potential. The resting potential maintains a concentration gradient across the membrane, so sodium ions diffuse across the membrane into the cell and potassium ions diffuse out. After the action potential, these ions are pumped by active transport back to their original positions.

Examiner's tip

The creation of action potentials and their transmission along the neurone is a complex process. Break it down into small steps and remember the sequence of events.

In the **transmission** of an action potential, as sodium ions enter the neurone they diffuse along inside the neurone. This produces a local current inside the neurone, which alters the potential difference across the membrane and causes sodium ion channels further along the neurone to open. This allows sodium ions to enter further along the neurone and the action potential moves along the neurone (Figure 2.2).

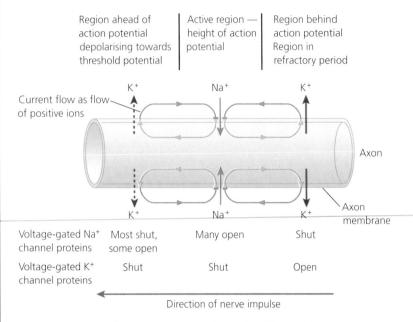

	Region ahead of action potential depolarising towards threshold potential	Active region — height of action potential	Region behind action potential Region in refractory period
Voltage-gated Na⁺ channel proteins	Most shut, some open	Many open	Shut
Voltage-gated K⁺ channel proteins	Shut	Shut	Open

Direction of nerve impulse

Figure 2.2 Current flow in a neurone in front of and behind the region with an action potential

After the potassium ions diffuse into the neurone, it looks as if the membrane is back to its resting potential. However, because the ions are effectively in the wrong places, the sodium–potassium ion pumps have to re-establish the correct positioning of the ions. Until this happens the axon enters a refractory period in which a new impulse cannot be generated.

Revision activity

Draw a series of diagrams to show the membrane at various stages of the action potential (resting potential, depolarisation and repolarisation). Your diagrams should show the activity of the sodium ion channels, the potassium ion channels and the sodium–potassium ion pumps.

Myelinated and non-myelinated neurones

Revised

Many neurones are **myelinated neurones**. This means that each individual neurone is surrounded by a fatty **myelin sheath** created by individual Schwann cells wrapped around the neurone. Between the Schwann cells are gaps called nodes of Ranvier, which occur at 1 mm intervals along the neurone. Ions cannot move across the membrane where the myelin sheath is in place — ion movements occur only at the nodes. In these neurones the local currents are stretched to carry the depolarisation from one node to the next. This speeds up transmission of the impulse as the action potential jumps between nodes. This is called salutatory conduction.

A **myelin sheath** is a non-conducting fatty layer around the neurone.

Non-myelinated neurones are still surrounded by Schwann cells. However, one Schwann cell surrounds several neurones and there are no nodes of Ranvier. In non-myelinated neurones, the action potential does not jump along the neurone.

Now test yourself

Tested

3 Explain how the action potential gets from one node of Ranvier to the next.

Answer on p. 122

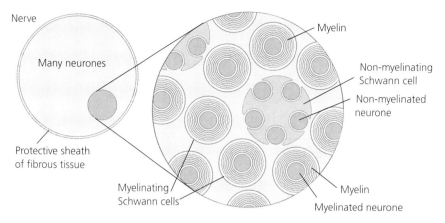

Figure 2.3 A section of a nerve as seen using a light microscope

Nervous communication

All neurones transmit action potentials in the same way. All action potentials are identical — this is known as the all-or-nothing rule. Therefore, there is no difference in the signals received by the brain from different stimuli. So how does the brain analyse the incoming information and interpret the stimulus so that we understand what the stimulus is and how intense it is? The neurones are connected into particular pathways. Neurones from one particular type of sensory receptor run to the same area of the brain. Therefore, when this part of the brain is stimulated, it is interpreted as that particular stimulus. The intensity of the stimulus affects the frequency at which the neurones fire — a higher intensity creates more frequent impulses in the neurones.

Transmission at synapses

The structure of a synapse

A **synapse** is a junction between two neurones. Neurones do not actually touch one another — there is a small gap between them called a synaptic cleft. Therefore, an action potential cannot pass directly from one neurone to another. The gap is bridged by the release of a chemical messenger called a **neurotransmitter**. Neurones communicate by cell signalling.

Neurones are well adapted to cell signalling:

- The first neurone ends in a bulge called the pre-synaptic knob or pre-synaptic bulb. This provides space to store the neurotransmitter in vesicles. It also provides a large surface area for the release of neurotransmitters such as **acetylcholine**.
- The second neurone contains many specialised proteins in its membrane. These proteins perform important roles:
  - They act as acetylcholine receptor sites (these sites have a shape that is complementary to the shape of the acetylcholine molecule).
  - They form sodium ion channels that open in response to acetylcholine molecules.

A **neurotransmitter** is a chemical that diffuses across the gap or cleft between two neurones.

Acetylcholine is the neurotransmitter in cholinergic synapses.

Transmission across a synapse

Transmission across a **cholinergic synapse** involves the release of acetylcholine into the synaptic cleft and the detection of that acetylcholine on the post-synaptic membrane (cell signalling). An action potential travelling along the pre-synaptic neurone reaches the pre-synaptic knob. Here it causes calcium ion channels to open. Calcium ions diffuse into the knob and cause the vesicles of acetylcholine to move towards and fuse with the pre-synaptic membrane. This releases the acetylcholine molecules into the cleft (exocytosis). The acetylcholine molecules diffuse across the cleft and bind to the acetylcholine receptor sites on the post-synaptic membrane. This opens the sodium ion channels in the post-synaptic membrane and sodium ions in the cleft diffuse into the post-synaptic neurone, causing depolarisation of the membrane. The enzyme acetylcholinesterase breaks down the acetylcholine, which ensures that the sodium ion channels close again. The choline is then recycled back into the pre-synaptic knob to make more acetylcholine.

Examiner's tip

The specification states 'Outline the role of neurotransmitters...', which means that little detail is required.

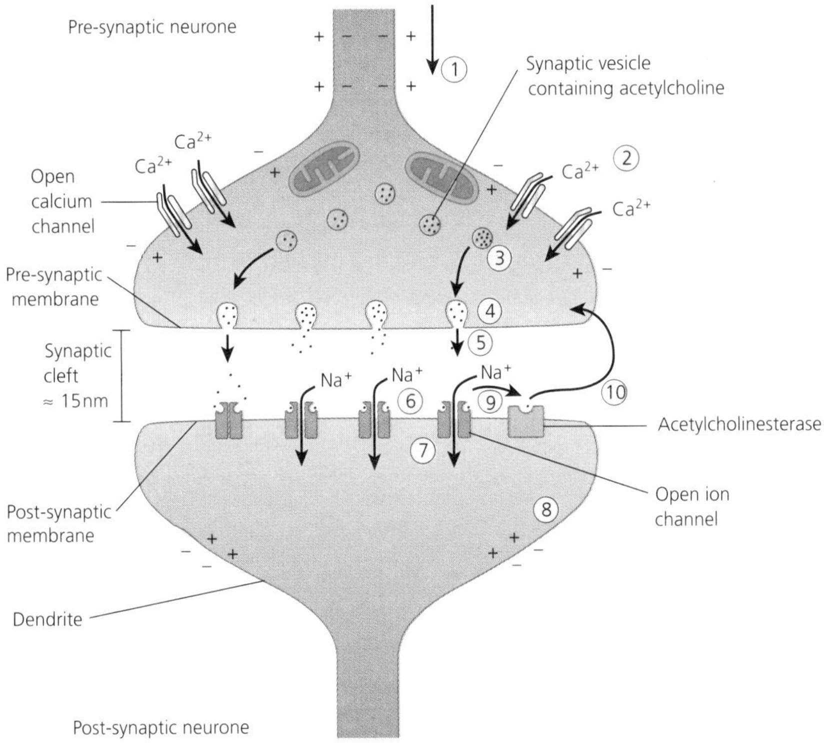

Typical mistake

Many candidates write a description that makes it sound as if the vesicles are released and diffuse across the cleft. However, the vesicles fuse to the membrane and release the acetylcholine molecules.

Revision activity

Draw a diagram of a synapse and annotate it to show how transmission occurs across the synapse.

Figure 2.4 The events that occur during the transmission of an impulse across a cholinergic synapse

The role of synapses

Synapses join neurones together, allowing transmission of a signal from one neurone to another (cell signalling). Action potentials travelling in the wrong direction can be stopped as there are no vesicles of acetylcholine in the post-synaptic neurone. Impulses from low-intensity stimuli can be filtered out so that there is no unnecessary response. This is achieved because many vesicles must be released to cause a post-synaptic action potential. Continuous unimportant stimuli can be ignored as the vesicles of acetylcholine run out after a while. This is called fatigue and prevents continuous responses to

Examiner's tip

This is another 'outline' statement in the specification, so little detail is needed.

continuous unimportant stimuli. This allows a form of behaviour called acclimatisation.

Action potentials in one neurone can be used to stimulate several post-synaptic neurones so that multiple responses can be achieved from one stimulus. Nerve impulses from more than one stimulus can be combined to create the same response — this can be used to magnify or amplify the response to a low-intensity stimulus. This is called summation. Inhibitory synapses can prevent the formation of an action potential in the post-synaptic neurone. Synapses also allow the formation of specialised nervous pathways, which are the basis of memory.

Exam practice

1   (a) (i)   Identify the molecules that prevent the movement of charged particles across a membrane.   [1]

    (ii) Describe how the membrane of a motor neurone is specialised to allow the movement of charged particles.   [3]

    (b) Small ions such as sodium and potassium leak across membranes. Explain how a neurone maintains the resting potential across its cell surface membrane despite this leakage.   [3]

    (c) Look at Figure 2.1 on p. 13. Use the information in the diagram to describe how an action potential is created. In your answer you should use appropriate technical terms, spelt correctly.   [5]

2   (a) (i)   The pre-synaptic knob contains many organelles. Suggest three organelles that may be found in unusually high numbers.   [3]

    (ii) State the function of each of these organelles.   [3]

    (b) (i)   Name the neurotransmitter used in cholinergic synapses.   [1]

    (ii) Describe the mechanism by which this neurotransmitter is released.   [4]

    (c) Modern headphones are available with sound cancelling technology. This works by emitting a constant frequency behind the music being played. Suggest how this could work to cancel other background sounds.   [2]

3   Complete the following paragraph.   [9]

    A synapse is a junction between two .... The action potential in the pre-synaptic neurone causes the release of .... from vesicles in the pre-synaptic knob. The neurotransmitter molecules .... across the synaptic cleft and bind to .... molecules in the post-synaptic membrane. These molecules have a specific shape that is .... to the shape of the neurotransmitter molecule. The binding of the neurotransmitter molecules opens .... ion channels and these ions flow .... the post-synaptic neurone. This .... the post-synaptic membrane, producing a new .... potential.

Answers and quick quiz 2 online

Online

Examiner's summary

By the end of this chapter you should be able to:

✔ Outline the roles of sensory receptors in mammals.

✔ Describe the structure and functions of sensory and motor neurones.

✔ Describe and explain how the resting potential is established and maintained.

✔ Describe and explain how an action potential is generated and transmitted in a myelinated neurone.

✔ Interpret graphs of the voltage changes taking place during the generation and transmission of an action potential.

✔ Outline the significance of the frequency of impulse transmission.

✔ Compare and contrast the structure and function of myelinated and non-myelinated neurones.

✔ Describe the structure of a cholinergic synapse.

✔ Outline the role of neurotransmitters in the transmission of action potentials.

✔ Outline the roles of synapses in the nervous system.

3 Hormones

Hormone action

The endocrine system   Revised

The endocrine system is a group of glands that release **hormones** directly into the blood. These glands are not linked other than by the blood circulatory system and the hormones that are transported in the blood. **Endocrine glands** are glands that have no ducts and secrete hormones directly into the blood.

Exocrine glands have ducts. The cells of the exocrine glands are usually arranged in small groups or acini that secrete molecules into a small duct or ductule. The ductules lead into larger ducts and eventually to a single duct that carries the product to where it is needed.

> **Hormones** are chemical signalling molecules that are released directly into the blood.

Revision activity

Produce a table to compare and contrast endocrine and exocrine glands.

Hormones   Revised

Hormones are complex molecules that act as chemical signals. They are released from an endocrine gland directly into the blood and are transported in the blood to their **target tissue** or organ. The target tissue is the site at which the hormone acts. The cells of the target tissue have cell surface receptors that have a complementary shape to the shape of the hormone.

Now test yourself

1 Explain why hormones are transmitted all over the body in the blood and yet affect only the correct target tissues.

Answer on p. 122

Tested

First and second messengers   Revised

The hormone is known as the **first messenger** as it carries the signal from the endocrine gland to the cells of the target tissue. The **second messenger** is a molecule inside the cells of the target tissue that transmits the signal from the membrane-bound receptor into the cell cytoplasm where an effect can be carried out. An example of a first and second messenger is the effect that **adrenaline** has on its target cells (Figure 3.1):

- Adrenaline in the blood binds to the membrane-bound receptor (an adrenergic receptor).
- A G protein is moved into place to activate the enzyme adenyl cyclase.
- The adenyl cyclase converts adenosine triphosphate (ATP) to **cyclic adenosine monophosphate (cAMP)**.
- The cAMP acts as a second messenger by moving into the cell cytoplasm and causing an effect in the cell — in this case, by activating specific enzymes.

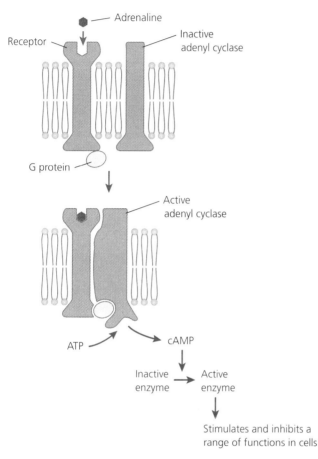

Figure 3.1 The role of the second messenger (cAMP) inside cells stimulated by adrenaline

Now test yourself

Tested

2 Suggest how cAMP can have different effects in different cells.

Answer on p. 122

The adrenal glands

Revised

The **adrenal glands** are endocrine glands that lie just above the kidneys. They have two regions: an outer cortex and an inner medulla. The cortex releases steroid-based hormones called adrenocorticoids. These contribute to the homeostatic maintenance of glucose and mineral concentrations in the blood, which also affect blood pressure. The adrenal medulla releases adrenaline, which prepares the body for activity. This includes the following effects:

- stimulates the breakdown of glycogen
- increases blood glucose concentration

Typical mistake

Candidates often forget that the adrenal glands have two regions and omit to describe the adrenal cortex and its hormones.

- increases heart rate
- increases blood flow to the muscles
- decreases blood flow to the gut and skin
- increases width of bronchioles to ease breathing
- increases blood pressure

Now test yourself   Tested

3   Write a brief explanation of how each listed effect of adrenaline helps to prepare the body for activity.

Answer on pp. 122–3

The pancreas

have ducts where are released the mol ↑   *no duct and release hormones directly into the blood*   Revised

The **pancreas** is both an exocrine gland and an endocrine gland (Figure 3.2).

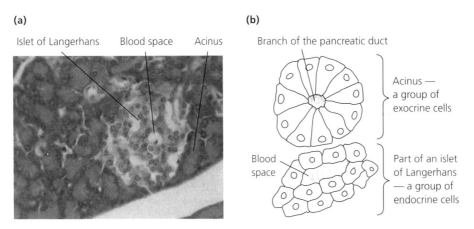

(a)

Islet of Langerhans   Blood space   Acinus

(b)

Branch of the pancreatic duct

Acinus — a group of exocrine cells

Blood space

Part of an islet of Langerhans — a group of endocrine cells

Figure 3.2 (a) A high-power photomicrograph of the endocrine and exocrine areas of the pancreas (b) A drawing showing cells from the two areas

Exocrine function

The majority of the pancreas consists of enzyme-producing cells arranged into acini (singular: acinus). An acinus is a group of cells arranged around a tiny ductule. The cells manufacture digestive enzymes that are released into the ductule. Many ductules combine together to form the pancreatic duct. The pancreatic duct carries the enzymes into the small intestine, where digestion takes place.

Endocrine function

In between the exocrine tissue are patches of endocrine tissue. These are called **islets of Langerhans** and they consist of two types of cell, alpha (α) and beta (β) cells:

- The alpha cells manufacture and release **glucagon** directly into the blood. Glucagon has the effect of increasing blood glucose concentrations.
- The beta cells manufacture and release **insulin**. Insulin has the opposite effect — it decreases blood glucose concentrations.

> The **islets of Langerhans** are small patches of endocrine tissue in the pancreas.

Regulating blood glucose and insulin

Blood glucose concentration

Revised

Glucose concentration in the blood is maintained at about 90 mg per 100 cm³. This is the set point, but concentration varies depending on what is eaten and the level of activity.

If blood glucose concentration gets too high:

- the glucose acts as a stimulus to cause the release of insulin from the beta cells in the islets of Langerhans
- the insulin is released directly into the blood and transported around the body — its target organ is the liver
- in the liver, the insulin binds to membrane-bound receptors on the liver cells
- the insulin causes the following effects in the liver cells: they absorb more glucose from the blood, convert the glucose to glycogen for storage, increase the use of glucose in respiration and inhibit the conversion of fats and glycogen to glucose
- insulin also affects the cells in muscles and fatty tissue. These cells increase their absorption of glucose from the blood and convert it to fat or glycogen

If blood glucose concentration gets too low:

- the beta cells stop producing insulin
- the alpha cells in the islets of Langerhans are stimulated to release glucagon directly into the blood
- glucagon is transported to the liver cells and binds to specific membrane-bound receptors
- these activate adenyl cyclase to produce cAMP, which acts as a second messenger inside the liver cells
- glucagon has two effects in the liver cells: it stimulates the conversion of glycogen to glucose and increases the conversion of fats and amino acids to glucose

Examiner's tip

In your written responses, you must be specific — refer to 'blood glucose concentration' rather than 'sugar concentration' or 'level of glucose'. Also make sure that all terms such as *glycogen*, *glucagon* and *glucose* are spelt correctly.

Typical mistake

Candidates do not learn the names of the substances well enough and often give spellings that suggest they are hedging their bets with the spellings, e.g. 'glucagen' or 'glycogon'.

Revision activity

Draw a flow diagram to show the response to blood glucose concentration getting too high and too low — start with the set point as a horizontal line and show an increase in concentration above the line and the sequence of events to bring the concentration back to set point, and repeat for a drop in concentration below the line.

Insulin

Figure 3.3 shows how the release of insulin from beta cells is controlled.

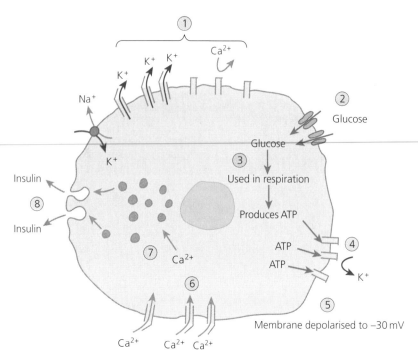

Figure 3.3 The release of insulin from beta cells in islet tissue in the pancreas

1 At normal concentrations of glucose, the beta cells maintain a membrane potential of –60–70 mV inside. There are open potassium ion channels in the membrane that allow the diffusion of potassium ions out of the cell.

2 High blood glucose concentrations cause glucose molecules to diffuse into the beta cells.

3 This glucose is then used in respiration to manufacture ATP.

4 The potassium ion channels respond to an increase in ATP concentration by closing. This stops potassium ions leaving the cell.

5 The concentration of potassium ions inside the cell increases as sodium–potassium ion pumps move potassium ions into the cell, causing partial depolarisation of the membrane.

6 As the membrane potential rises from –60 mV to –30 mV, the calcium ion channels respond by opening.

7 These calcium ions diffuse into the beta cells and cause vesicles containing insulin to move to, and fuse with, the plasma membrane.

8 Insulin is released from the cell by exocytosis.

Diabetes and its treatment

Diabetes mellitus is the inability to control blood glucose concentration. There are two types of diabetes mellitus, **Type 1 (insulin-dependent)** and **Type 2 (non-insulin-dependent)**, as shown in Table 3.1.

Table 3.1 Comparing Type 1 and Type 2 diabetes

Feature	Type 1 diabetes	Type 2 diabetes
Uncontrolled blood glucose	Yes	Yes
Insulin-dependent	Yes	No
Onset	Usually juvenile	Usually in middle age
Cause	Inactive beta cells, possibly caused by an autoimmune response or a virus	Liver cells become less responsive to insulin. Linked to obesity, high levels of refined sugars in the diet and family history. More common in people of Asian or African-Caribbean origin
Treated by insulin injections	Yes	No
Treated by careful monitoring of the diet and regular exercise	Yes	Yes

Tested ☐

Now test yourself

5   Explain why someone with diabetes mellitus may feel tired and unable to exercise strenuously or for long periods.

Answer on p. 123

Insulin for injection to treat Type 1 diabetes used to be isolated from the pancreases of pigs slaughtered for food. **Genetically modified bacteria** can now be used to manufacture human insulin, which has a number of advantages:

● It is human insulin rather than pig insulin, therefore it is more effective.

● There is less chance of rejection.

● It is cheaper to produce.

● It is easier to increase production if demand increases.

● There are fewer ethical or moral objections.

Research is continuing into the possibility of using **stem cells** to grow new beta cells in the pancreas of patients with Type 1 diabetes. This would be a permanent cure and free diabetes patients from the need to inject insulin every day.

Examiner's tip

The use of genetically modified bacteria to produce insulin is an example of how scientific advances benefit society — one of the aspects of How Science Works. Don't forget that How Science Works is tested in every examination paper.

Examiner's tip

The potential use of stem cells is another aspect of How Science Works — questions are likely to test your ability to make use of the information given in the question.

Control of heart rate

Revised ☐

Heart rate is controlled by both **hormonal mechanisms** and **nervous mechanisms**. The heart muscle is myogenic, which means it can generate its own rhythm. However, this rhythm is overridden by the sinoatrial node (SAN). The SAN initiates waves of excitation that cause the heart to beat. The rate at which the SAN initiates contractions is affected by two nerves from the cardiovascular centre in the medulla of the brain (Figure 3.4):

● The decelerator nerve (vagus nerve) releases acetylcholine, which reduces the heart rate.

● The accelerator nerve releases noradrenaline, which is similar to adrenaline and increases the heart rate.

These nerves also affect the atrioventricular node (AVN). The heart also responds to hormones in the blood, particularly adrenaline.

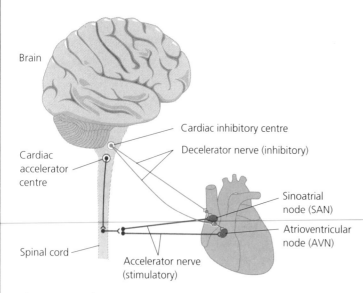

Brain

Cardiac inhibitory centre

Decelerator nerve (inhibitory)

Cardiac accelerator centre

Sinoatrial node (SAN)

Atrioventricular node (AVN)

Spinal cord

Accelerator nerve (stimulatory)

Figure 3.4 Dual innervation of the heart by decelerator and accelerator nerves

Exam practice

1 **(a)** Describe the difference between a first messenger and a second messenger. [2]

(b) Suggest why hormones, which are polypeptides, need to act through a receptor and a second messenger. [3]

(c) Explain how the pancreas can be both an endocrine gland and an exocrine gland. [3]

2 **(a)** Describe the regulation of blood glucose concentration. In your answer you should use appropriate technical terms, spelt correctly. [5]

(b) Describe how the beta cells in the pancreas respond to high blood glucose concentration. [3]

(c) Glucokinase is an enzyme that facilitates the phosphorylation of glucose to glucose-6-phosphate. It occurs in the cells of the pancreas. It plays an important role in the regulation of carbohydrate metabolism by acting as a glucose sensor, triggering shifts in metabolism or cell function in response to rising or falling levels of glucose. Mutations of the gene for this enzyme can cause unusual forms of diabetes. Suggest how a modified version of this enzyme could cause symptoms of diabetes. [2]

3 **(a)** Distinguish between Type 1 and Type 2 diabetes. [3]

(b) Suggest what advice might be given to a person with Type 2 diabetes. [3]

(c) State the potential advantages of treating Type 1 diabetes with stem cells. [2]

Answers and quick quiz 3 online

Online

Examiner's summary

By the end of this chapter you should be able to:

✔ Define the terms *endocrine gland, exocrine gland, hormone* and *target tissue*.

✔ Explain the meaning of the terms *first messenger* and *second messenger*.

✔ Describe the function of the adrenal glands.

✔ Describe the histology of the pancreas and outline its role as an exocrine and an endocrine organ.

✔ Explain how blood glucose concentration is regulated.

✔ Outline how insulin secretion from beta cells is controlled.

✔ Compare and contrast the causes of Type 1 and Type 2 diabetes mellitus.

✔ Discuss the use of insulin from genetically modified bacteria and the potential use of stem cells to treat diabetes mellitus.

✔ Outline how heart rate is controlled.

4 Excretion

Removing metabolic wastes

Examiner's tip

This examination paper includes synoptic marks. These test your:
- understanding of the principles behind different processes
- ability to make links back to other parts of the specification

The obvious links here are:
- the structure and ventilation of the lungs, the circulatory system, osmosis, active transport and diffusion from F211
- the structure of amino acids and the effect that changes in pH can have on protein structure and enzyme action from F212

Excretion
Revised ☐

Excretion is the removal of **metabolic wastes** from the body, which are waste products from processes that have occurred inside the cells. Metabolic wastes include:

- **carbon dioxide**, which is removed via the lungs
- **nitrogenous waste**, which is removed via the kidneys

Removal of carbon dioxide
The removal of carbon dioxide involves a sequence of steps:

1 Carbon dioxide is removed from cells by diffusion.

2 It enters the blood, which transports it to the alveoli of the lungs.

3 Through gaseous exchange, the carbon dioxide diffuses into the air spaces of the alveoli.

4 Ventilation expels the carbon dioxide from the lungs.

If excess carbon dioxide builds up, it becomes toxic. Carbon dioxide reacts with water to form carbonic acid. A build-up in acidity can affect the activity of enzymes. Some carbon dioxide binds directly to haemoglobin and most is transported by a mechanism that involves haemoglobin, so less haemoglobin will be available for oxygen transport.

Removal of nitrogenous waste
Nitrogenous waste includes ammonia, **urea** and uric acid. Ammonia is produced by the deamination of excess amino acids. It is highly toxic and increases the pH of cell cytoplasm, affecting the activity of enzymes. It is converted in the liver to less toxic urea, which is then removed from the blood in the kidneys to form part of urine.

Typical mistake

Do not confuse excretion with egestion, which is the elimination of undigested food.

Examiner's tip

This is an obvious area for synoptic testing of material from your AS biology.

Revision activity

Draw a diagram of a red blood cell and annotate it to show how carbon dioxide is transported.

The liver

The **liver** consists of cells called hepatocytes. These are all the same and are metabolically very active. They contain a wide range of enzymes and organelles to help process substances in the blood. The structure of the liver has evolved to deliver as much blood as possible to the hepatocytes. The liver receives blood from two sources:

- The hepatic artery carries oxygenated blood from the aorta.
- The hepatic portal vein carries deoxygenated blood from the digestive system.

Inside the liver, these vessels divide up to form narrow branches that flow along the portal areas between the many liver lobules (Figure 4.1). Each lobule consists of many hepatocytes arranged in columns radiating out from the centre of the lobule like the spokes of a wheel. The blood from both sources mixes as it enters a lobule and flows along a narrow channel called a sinusoid. This channel runs between the hepatocytes towards the centre of the lobule. As blood flows past the hepatocytes, exchange occurs across the cell surface membranes. Many substances are removed from the blood to be processed in the hepatocytes, whereas other substances such as glucose may be released into the blood. At the centre of each lobule the sinusoids flow into a branch of the hepatic vein, which carries blood back out of the liver.

Bile canaliculi (singular: canaliculus) also run between the columns of hepatocytes. These carry bile in the opposite direction to the sinusoids. The bile transports certain substances towards the bile duct, which lies in the portal area between adjacent lobules. These substances include:

- bile salts, which are used in digestion
- bile pigments, which are excretory products

(a)

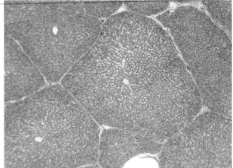

(b)

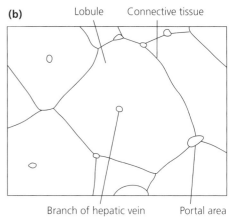

Figure 4.1 (a) A low-power photomicrograph of some liver lobules (b) A plan drawing made from the photomicrograph

Typical mistake

Many candidates confuse the intralobular vessel (the hepatic vein) with the interlobular vessels in the portal area (the hepatic artery and the hepatic portal vein).

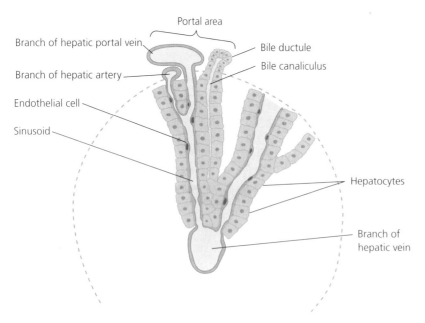

Figure 4.2 The arrangement of blood vessels, sinusoids, hepatocytes and bile canaliculi inside each liver lobule

Urea formation Revised

Excess amino acids are **deaminated** in the hepatocytes, which involves the removal of the amino group to form ammonia. It leaves an organic acid residue that can be used in respiration:

amino acid → ammonia + organic acid residue

> **Deamination** is the removal of the amino group from an amino acid.

The ammonia is highly toxic and must be quickly converted to the less toxic nitrogenous compound urea:

ammonia + carbon dioxide → urea

This is achieved in the **ornithine cycle** (Figure 4.3).

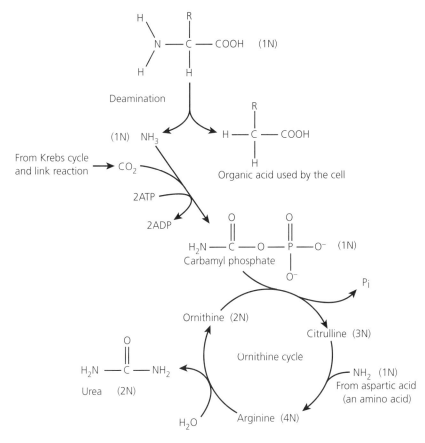

Figure 4.3 Deamination and the ornithine cycle

> **Examiner's tip**
>
> Remember the details of the ornithine cycle as some questions do ask for quite a bit of detail.

> **Now test yourself**
>
> 3 Explain why mammals must convert ammonia to urea whereas fish can excrete ammonia directly into their surroundings.
>
> Answer on p. 123
>
> Tested

Detoxification Revised

The liver has a number of other roles in metabolism, including:

- storing glycogen
- synthesis of proteins, cholesterol and bile salts
- transamination
- **detoxification** — in particular, the detoxification of hydrogen peroxide, alcohol (which is oxidised to ethanal and then to ethanoic acid before being converted to acetyl coenzyme A) and other drugs such as paracetamol, steroids, antibiotics

> **Revision activity**
>
> Sketch a liver cell and annotate it with arrows to show the movement of substances into and out of the cell. Don't forget the knowledge you have gained in Chapter 3.

The kidney

The kidney is supplied with blood from the renal artery. Inside the kidney, the blood passes into specialised capillary beds (**glomeruli**), which filter the blood before it returns to the body via the renal vein. Waste products in the blood enter the ureter to flow to the bladder. The kidney consists of two layers: the outer cortex (Figure 4.5) and inner medulla (Figure 4.6). In the centre is the pelvis, which collects the urine produced and leads into the ureter.

> **Glomeruli** (singular: glomerulus) are tiny knots of capillaries at the start of each nephron.
>
> A **nephron** is a small kidney tubule that filters the blood and produces urine.

Each kidney contains around 1 million tiny tubules called **nephrons**. These start at a knot of capillaries called the glomerulus in the cortex. They follow a convoluted path into the collecting ducts that lead into the pelvis.

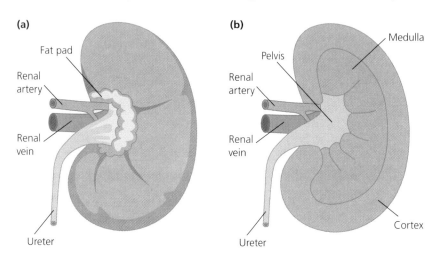

Figure 4.4 (a) The gross structure of a kidney: external view (b) A vertical section

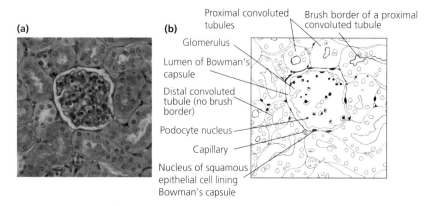

Figure 4.5 (a) A photomicrograph of part of the cortex of the kidney (b) A drawing of part of the photomicrograph

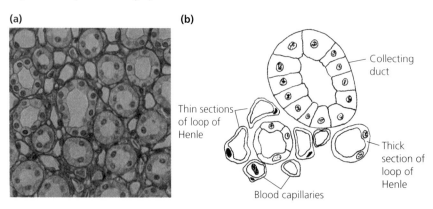

Figure 4.6 (a) A photomicrograph of part of the medulla of the kidney (b) A drawing of part of the photomicrograph

Nephrons

Revised

Each nephron (Figure 4.7) starts in the cortex at a glomerulus, which is where ultrafiltration takes place. Surrounding the glomerulus is a cup-shaped structure called the Bowman's capsule. Fluid passes from the capsule to the proximal convoluted tubule where **selective reabsorption** takes place. The fluid then passes along the **loop of Henle** down into the medulla and back out to the cortex. From the loop of Henle, the fluid passes into the distal convoluted tubule and into the collecting duct where reabsorption of water takes place.

> The **loop of Henle** is an extended loop of the nephron that runs into the medulla of the kidney.

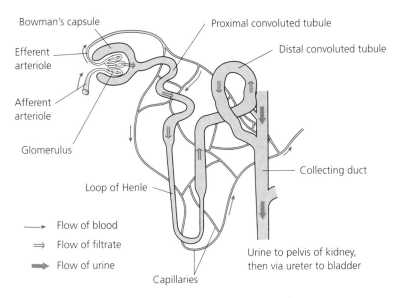

Figure 4.7 A kidney nephron and associated blood vessels

Ultrafiltration

Revised

Ultrafiltration (Figure 4.8) is the filtering of the blood through a basement membrane which acts as a fine sieve. The arteriole leading into the glomerulus (the afferent arteriole) is wider than the arteriole leading out (the efferent arteriole), which maintains the blood pressure in the capillaries. Blood plasma containing the dissolved components of blood is squeezed out of the capillaries.

The Bowman's capsule surrounding the glomerulus has a specialised inner layer of cells called **podocytes**. These have major (primary) processes and minor (secondary) processes, which leave small gaps between the cells that allow the fluid to enter the capsule. Between the capillary of the glomerulus and the podocytes is a basement membrane. This is the filter and it allows the passage of molecules with a relative molecular mass of 69 000 or less.

> **Podocytes** are specialised cells lining the Bowman's capsule that leave gaps between them to reduce the resistance to flow of fluid.

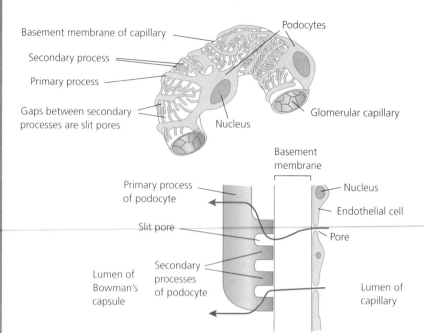

Figure 4.8 Ultrafiltration in the glomerulus

Selective reabsorption

Selective reabsorption (Figure 4.9) is the removal of certain substances from the fluid in the tubule, leaving other substances. The cells of the proximal convoluted tubule are specialised to reabsorb glucose, amino acids and mineral ions from the fluid in the tubule. The membranes of these cells contain sodium–potassium ion pumps and cotransport proteins. The membranes are also folded into microvilli to increase their surface area. The cells also contain many mitochondria to supply ATP for active transport. Sodium ions are pumped out of the cell through the basal membrane towards the blood capillary. This is active transport and requires ATP. This reduces the concentration of sodium ions inside the cell, creating a concentration gradient. Sodium ions diffuse into the cell from the fluid in the tubule, but they diffuse through cotransport proteins that bring glucose molecules (or amino acids) into the cell at the same time as the sodium ions. The concentration of glucose in the cell rises and it can diffuse out through the basal membrane into the blood capillary.

Typical mistake

Many candidates confuse an 'active process' with 'active transport'. The reabsorption of glucose is an active process that includes the active transport of sodium ions out of the cells lining the tubule.

Now test yourself

Tested

4 Explain why it is not essential to reabsorb proteins from the fluid in the nephron.

Answer on p. 123

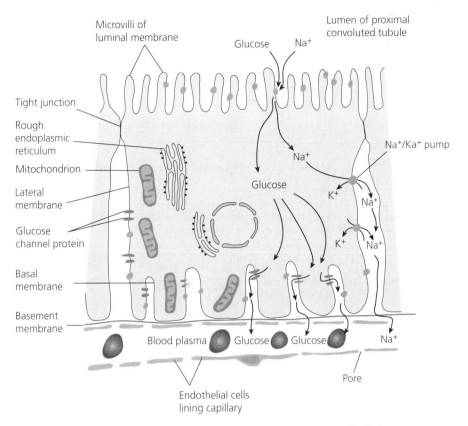

Figure 4.9 Selective reabsorption by a cell of the proximal convoluted tubule

The loop of Henle

The loop of Henle modifies the fluid in the tubule to help conserve water.

1 As the fluid passes down the loop, it becomes more concentrated due to salts diffusing into the fluid from the surrounding tissue.

2 As the fluid moves up the loop, it becomes more dilute as salts are pumped out of the fluid in the tubule.

3 This passes salts from the ascending limb to the descending limb so that the fluid becomes increasingly concentrated. It is called a **hairpin countercurrent multiplier**.

4 The effect is to create a concentration gradient in the medulla of the kidney. The salt becomes increasingly concentrated towards the centre of the kidney.

5 This means that the water potential of the tissue fluid is very low.

6 As fluid passes along the collecting duct from the cortex to the pelvis, water is withdrawn by osmosis into the tissue fluid.

7 This results in urine that has a lower water potential than blood and conserves water.

> A **hairpin countercurrent multiplier** is a mechanism to increase the concentration of substances in a tube where substances are passed between two parts of the same tube in which the fluid flows in opposite directions.

Typical mistake

Many candidates get confused with water potentials and concentrations. Don't use the term *water concentration* and remember that a higher salt concentration causes a lower water potential.

Examiner's tip

The way in which the loop of Henle works is complicated — remember that its role is to produce a salt concentration gradient in the medulla. This is another excellent area in which the examiner can test your knowledge of AS material such as osmosis.

Control of water potential

Osmoreceptors and the posterior pituitary gland

Osmoreceptors in the **hypothalamus** monitor the water potential of the blood. When blood water potential falls too low, water is withdrawn from the osmoreceptor cells via osmosis and the cells shrink. This shrinkage causes the osmoreceptors to stimulate the neurosecretory cells that lead down into the **posterior pituitary gland**. The neurosecretory cells release antidiuretic hormone (ADH) directly into the blood. The ADH travels around the body, but its target tissue is the collecting ducts in the kidney. It binds to cell surface receptors on the cells lining the collecting ducts and causes the release of cAMP inside the cell as the second messenger. The end result is the insertion of more water-permeable channels into the membrane lining the lumen of the collecting duct. This makes the cells more permeable to water. More water is reabsorbed from the urine in the collecting duct and the blood water potential rises again. If the blood water potential rises too high, the osmoreceptor cells swell and less ADH is released.

> **Examiner's tip**
>
> This is an excellent example of negative feedback.

> **Now test yourself**
>
> 5 Explain why the mechanism involving osmoreceptors in the hypothalamus is called negative feedback.
>
> Answer on p. 123
>
> Tested

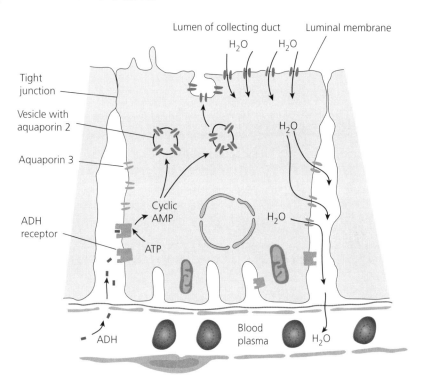

Tight junction

Vesicle with aquaporin 2

Aquaporin 3

ADH receptor

Lumen of collecting duct   Luminal membrane

H_2O H_2O

H_2O

Cyclic AMP

H_2O

ATP

H_2O

ADH

Blood plasma H_2O

Figure 4.10 The changes that occur in a cell in a collecting duct in response to stimulation by ADH (aquaporin 1 is found in the cell surface membranes of the proximal convoluted tubule and the descending limb of the loop of Henle)

> **Revision activity**
>
> Draw a flow diagram to show the response to blood water potential getting too high and too low. Start with the set point as a horizontal line above the line and show an increase in water potential above the line and the sequence of events to bring the water potential back to set point, and repeat for a drop in water potential below the line.

Renal failure and urine samples

Kidney failure

Revised

Kidney failure can be fatal as metabolic wastes such as urea, toxins and excess salt and water are not excreted. Kidney failure can be treated by:

- **renal dialysis** — this involves connection to a dialysis machine which takes blood from a vein in the arm and passes the blood over a dialysing membrane. The membrane is partially permeable and exchanges materials between the blood and special dialysis fluid. The fluid contains all the correct concentrations of substances in the blood and must be constantly refreshed. Over time the unwanted substances in the blood, such as urea, diffuse into the dialysing fluid and are removed. The blood is returned to a vein in the arm. Peritoneal dialysis is an alternative in which dialysing fluid is placed in the abdominal cavity for exchange to occur across the peritoneum. The disadvantage of dialysis is that the patient needs to spend many hours a week connected to the machine and they also have to monitor dietary intake carefully.

- **transplant** — kidneys can be transplanted successfully and the patient can then live a normal life. However, care must be taken to ensure a good match between the donor and recipient.

Revision activity

Draw a mind map to link all the possible consequences of kidney failure.

Urine tests

Revised

Drugs and hormones are treated by the liver and will be excreted by the kidneys in urine. Urine can be tested for the presence of hormones, drugs or traces of drugs.

Pregnancy tests

Human embryos release a hormone called human chorionic gonadotrophin (hCG), which appears in the urine of the mother. **Pregnancy testing** sticks contain monoclonal antibodies specific to this hormone. The sticks have some antibodies that are attached to a tiny blue bead. These antibodies are free to move. Other antibodies are attached (immobilised) in a line across the stick. If the hormone is present, it binds to the free antibodies and to the immobilised antibodies holding the beads in a line across the stick.

Anabolic steroid tests

Some athletes are tempted to use **anabolic steroids** to boost their muscle growth and strength. Urine can be tested by gas chromatography in which a gaseous solvent is used to separate substances in the urine or by mass spectrometry. Use of anabolic steroids gives an unfair advantage and has been banned in sport.

Exam practice

1 (a) List three functions of the hepatocytes. [3]

(b) Describe the arrangement of hepatocytes in the liver and explain how this ensures that as much blood as possible can be processed by them. [4]

2 The following diagram shows the pathways in which alcohol (ethanol) and fatty acids are broken down into acetyl coenzyme A.

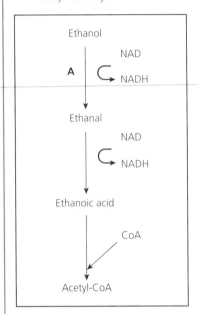
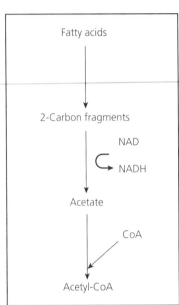

(a) Name the type of reaction occurring at step A. [1]

(b) State two similarities in these two reaction pathways. [2]

(c) Drinking too much alcohol can lead to a condition known as a fatty liver in which the liver swells and accumulates fats. Suggest why this occurs. [3]

3 The following table shows the ratio of concentrations of substances found in the capillaries relative to the concentration found in the Bowman's capsule.

Substance	Concentration in capillary relative to Bowman's capsule
Protein	15 000 : 1
Amino acid	1 : 1
Glucose	1 : 1
Urea	1 : 1
Sodium ions	1 : 1

(a) Name the process that occurs between the capillary and the Bowman's capsule. [1]

(b) Identify one substance that cannot enter the Bowman's capsule. [1]

(c) Describe what prevents this substance entering the Bowman's capsule. [2]

(d) Explain why the concentration of amino acids in the Bowman's capsule is the same as that in the capillary. [2]

4 (a) Name the region of the nephron where selective reabsorption takes place. [1]

(b) Describe how selective reabsorption of amino acids is achieved. In your answer you should use appropriate technical terms, spelt correctly. [5]

(c) Name one other substance that is selectively reabsorbed. [1]

(d) Describe three features of the cells of this region of the nephron that adapt them to their role of selective reabsorption and explain how each adaptation helps the process. [6]

Answers and quick quiz 4 online

Online

Examiner's summary

By the end of this chapter you should be able to:

✔ Define the term *excretion*.

✔ Explain the importance of removing metabolic wastes from the body.

✔ Describe the histology and gross structure of the liver.

✔ Describe the formation of urea in the liver including the ornithine cycle.

✔ Describe the roles of the liver in detoxification.

✔ Describe the histology and gross structure of the kidney.

✔ Describe the detailed structure of a nephron and its associated blood vessels.

✔ Describe and explain the production of urine.

✔ Explain the control of the water content of the blood.

✔ Outline the problems arising from kidney failure and how kidney failure can be treated.

✔ Describe how urine samples can be tested for pregnancy and misuse of anabolic steroids.

5 Photosynthesis

The mechanisms of photosynthesis

Autotrophs and heterotrophs   Revised ☐

Autotrophs are organisms that absorb small inorganic substances and convert them into large organic molecules. A source of energy is required and in most autotrophs this is light from the sun. **Light** is used in **photosynthesis** as a source of energy to produce **complex organic molecules**.

Heterotrophs are organisms that obtain their energy from large organic molecules. These molecules are digested either externally (in the case of fungi) or internally (in the case of animals) and absorbed into the body.

An **autotroph** is an organism that absorbs inorganic substances and converts them into complex organic molecules.

A **heterotroph** is an organism that obtains energy from large organic molecules.

Respiration   Revised ☐

The organic molecules (such as glucose) made by autotrophs or consumed by heterotrophs are used in respiration. Respiration converts the complex organic molecules into simple inorganic molecules (carbon dioxide and water), releasing energy that can be used by the organism. Respiration in plants and animals depends on two products of photosynthesis:

- The complex organic molecules that are broken down in respiration are formed from the products of photosynthesis.

- The oxygen released as a by-product in photosynthesis enables the complex organic molecules to be fully broken down in aerobic respiration.

Typical mistake

Many candidates forget that plants still need to respire — they photosynthesise during the day but they respire all the time.

Chloroplasts   Revised ☐

Photosynthesis takes place in **chloroplasts**, which are specialised organelles found inside plant cells (Figure 5.1). It has two stages, known as the **light-dependent stage** and the **light-independent stage**. The structure of a chloroplast is well adapted to the process of photosynthesis:

- The chloroplast has two membranes: inner and outer. Together they form the chloroplast envelope.

- The chloroplast contains many **thylakoids** that are arranged to form stacks of discs called grana (singular: granum). These produce a large surface area for the reactions associated with the light-dependent stage.
- The thylakoids contain many photosynthetic pigments arranged with enzymes and coenzymes to enable the light-dependent stage to proceed effectively in the thylakoid membranes.
- The grana are surrounded by a fluid called **stroma**, which contains all the enzymes needed for the light-independent stage of photosynthesis.

> **Thylakoids** are membrane-bound compartments in the chloroplast. They consist of a thylakoid membrane surrounding a thylakoid lumen.
> **Stroma** is the fluid inside a chloroplast.

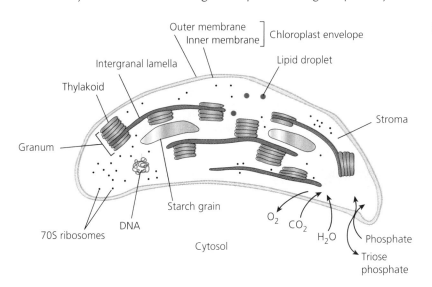

Figure 5.1 The structure of a chloroplast and the exchanges that occur with the cytosol

> **Revision activity**
> Sketch a diagram of a chloroplast and annotate it with information about the role of each structure in photosynthesis.

Photosynthetic pigments

Revised

Photosynthetic pigments are large organic molecules that absorb light and convert it into a form of energy that can be used in photosynthesis.

There is a range of photosynthetic pigments that each absorb a range of wavelengths. Each pigment has a specific peak of absorption at which it is most effective at absorbing light energy. Chlorophyll a is the main pigment, which absorbs red and blue light but reflects green light. It is a combination of a number of similar molecules, but chlorophyll a is known as the **primary pigment**. It is yellow-green and has two main forms: P_{680} and P_{700}. P_{680} has a peak of absorption at 680 nm whereas P_{700} has a peak of absorption at 700 nm.

Other photosynthetic pigments are called **accessory pigments**. These absorb additional wavelengths that are not absorbed well by chlorophyll a, and include:

- chlorophyll b, which is blue-green
- carotene, which looks orange as it reflects orange light and absorbs blue light
- xanthophyll, which looks yellow as it reflects yellow light and absorbs blue light

> **Photosynthetic pigments** are pigments that absorb light energy over a range of wavelengths.
> The **primary pigment** is the chlorophyll a found in the reaction centre.
> **Accessory pigments** absorb energy and pass it to the primary pigment.

> **Typical mistake**
> Many candidates seem to think that these two forms of chlorophyll a absorb light at only one wavelength, but don't forget that the figures refer to the peak of absorption.

Accessory pigments are arranged in a funnel-shaped light-harvesting apparatus in the thylakoid membranes. They absorb light and pass its energy down the funnel. Chlorophyll a is found at the base of the funnel in the reaction centre. Each molecule of chlorophyll contains a magnesium atom. Electrons in the magnesium are excited by light energy hitting the molecule or by energy passed on from the accessory pigments. The excited electrons are more energetic and move to higher energy orbits. The energy held by the electrons can then be used in the light-dependent stage of photosynthesis.

Now test yourself

Tested

1 Explain why chlorophyll is green.
2 Suggest why deciduous leaves turn yellow in autumn.

Answers on p. 123

Stage 1: Converting light energy to chemical energy

The light-dependent stage

Revised

The light-dependent stage of photosynthesis occurs only in the presence of light and in the thylakoid membranes. Each membrane contains a lot of light-harvesting funnels, each with a reaction centre at the base. There are two types of reaction centre, each containing a photosystem with a slightly different version of chlorophyll a:

● Photosystem I contains chlorophyll a (P_{700}).
● Photosystem II contains chlorophyll a (P_{680}).

The remaining enzymes, coenzymes and electron carriers are located in the membrane close to the reaction centres.

Photolysis

Photolysis means 'splitting by light'. One of the enzymes is used to split water to produce H^+ ions and electrons:

$$2H_2O \rightarrow 4H^+ + 4e^- + O_2$$

The hydrogen ions and electrons are used during the light-dependent stage. Because the products are taken away by the reactions of the light-dependent stage, this reaction is driven by light. Oxygen is released as a by-product. Photolysis occurs in association with photosystem II.

Cyclic photophosphorylation

Photophosphorylation is the combination of ADP and a phosphate group in the presence of light. **Cyclic photophosphorylation** uses the energy from an electron that has been excited by light absorption and returns the electron to its original position (Figure 5.2). In photosystem I (P_{700}), the energy from an electron excited by light can be used to generate **ATP**. The electron passes back to the same pigment system, which is why this process is called cyclic photophosphorylation:

1 The excited electron is picked up by an electron carrier.

2 It travels down an electron transport chain.

3 Energy is released in small amounts and used to pump hydrogen ions across the thylakoid membrane into the thylakoids.

4 This builds up a concentration gradient.

5 Hydrogen ions diffuse back out of the thylakoids through specialised channels attached to the enzyme **ATP synthase**.

6 The movement of the hydrogen ions provides the energy to combine ADP with P_i making ATP.

7 This process used to manufacture ATP is called chemiosmosis.

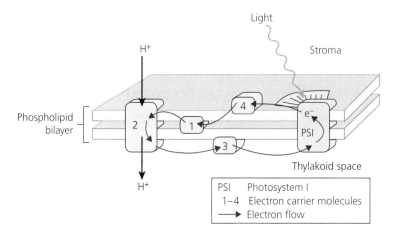

Figure 5.2 Cyclic photophosphorylation

Non-cyclic photophosphorylation

Non-cyclic photophosphorylation uses the energy from an electron that has been excited by light absorption, but does not return it to its original position: the excited electron from photosystem I may take another route. As the electron is not returned to photosystem I, this process is called non-cyclic photophosphorylation:

1 The electron from photosystem I is combined with H+ (from photolysis) and NADP to make **reduced NADP**. This leaves photosystem I unbalanced as it is missing an electron.

2 An electron from photosystem II (P_{680}) is also excited.

3 This electron is picked up by an electron carrier and taken down an electron transport chain to photosystem I. This replaces the electron lost from photosystem I.

4 The electron transport chain is also associated with the production of ATP through chemiosmosis.

5 However, photosystem II is left unbalanced as it is missing an electron. This is replaced by the electron released from photolysis.

The products of the light-dependent stage are ATP and reduced NADP. These are used in the light-independent stage. A by-product of photolysis is oxygen, which is released from the plant.

Examiner's tip

This topic seems complicated, but if you view the process as a flow diagram it is easy to follow. Remember that no biochemical detail is needed.

Revision activity

Draw a flow diagram of the reactions involved in the light-dependent stage (this diagram is often referred to as the Z-scheme).

Now test yourself

3 Explain why the reactions of the light-dependent stage are given the names cyclic and non-cyclic photophosphorylation.

Answer on p. 123

Tested

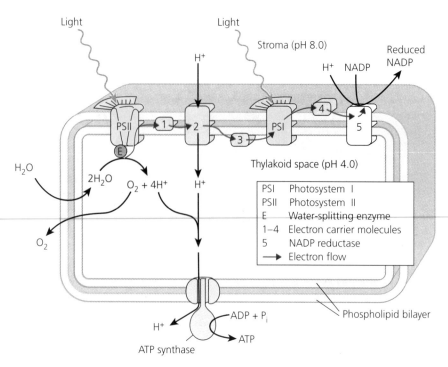

Figure 5.3 The light-dependent stage of photosynthesis

Stage 2: The Calvin cycle

The light-independent stage of photosynthesis can occur in the dark, but it uses the products of the light-dependent stage. It occurs in the stroma of the chloroplasts, which contains all the enzymes needed for fixing **carbon dioxide** to produce complex organic molecules.

Carbon dioxide is absorbed from the atmosphere. It is combined with **ribulose bisphosphate (RuBP)** in a process called fixing. The enzyme **ribulose bisphosphate carboxylase (rubisco)** speeds up this process. An unstable 6-carbon compound is produced, which rapidly breaks down into two molecules of **glycerate 3-phosphate (GP)**. This GP is then converted into **triose phosphate (TP)** by reduction using the reduced NADP and ATP from the light-dependent stage. The TP can then be used to manufacture the large organic molecules needed by the plant. These include:

- **carbohydrates** such as glucose
- **lipids**
- **amino acids**

However, most of the TP is **recycled** to produce more RuBP. This process requires more of the ATP made in the light-dependent stage to provide energy and a phosphate group.

Examiner's tip

Remember that no biochemical detail is needed.

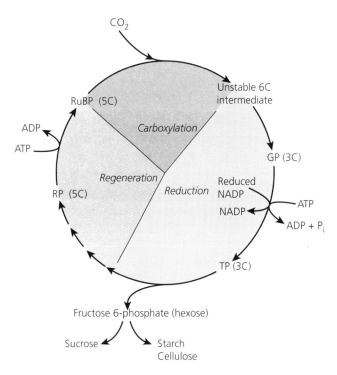

Figure 5.4 The Calvin cycle

Factors affecting the rate of photosynthesis

Limiting factors

Revised

Limiting factors are things that can limit the rate of a reaction if there is a limited supply. The rate of photosynthesis can be affected by a number of limiting factors. Only one factor can limit the rate of photosynthesis at any one time, so these factors interact with one another.

If **carbon dioxide concentration** rises, the rate of photosynthesis increases until another limiting factor prevents any further increase. If the concentration of carbon dioxide falls, this affects fixation. Less carbon dioxide can be fixed so less GP is made. This means that less TP and less glucose are made. The concentration of RuBP will rise.

If **light intensity** increases, the rate of photosynthesis increases until another factor prevents any further increase. A reduction in light intensity slows the light-dependent stage. As a result, less ATP and reduced NADP are available to power the light-independent stage (Calvin cycle). Less GP is converted to TP and less TP can be converted to RuBP, so the concentration of GP remains high while the concentration of RuBP and TP drops. This means that less carbon dioxide can be fixed.

Temperature has little effect on the light-dependent stage. However, the light-independent stage is controlled by enzymes. A drop in temperature reduces the rate of enzyme-controlled reactions. An increase in temperature increases the rate of enzyme-controlled reactions until the enzymes are denatured.

Examiner's tip

Remember to refer to the concentration of carbon dioxide.

Typical mistake

Many candidates lose marks by referring to 'light' changing rather than 'light intensity' changing.

Now test yourself

4 Explain why reduced light intensity leads to an increase in the concentration of GP and a reduction in the amount of glucose produced.

Answer on p. 123

Tested

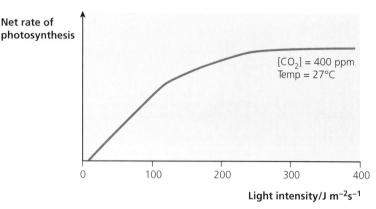

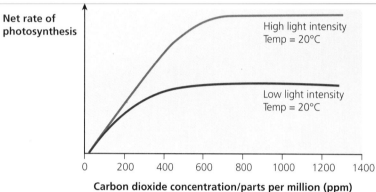

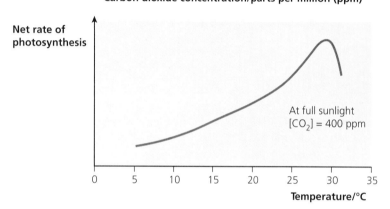

Figure 5.5 The effects of light intensity, carbon dioxide concentration and temperature on the rate of photosynthesis of common orache, *Atriplex patula*

Measuring the rate of photosynthesis

Revised

The easiest way to investigate the rate of photosynthesis is through the production of oxygen. It is possible to count bubbles of oxygen released from the cut stem of an aquatic plant such as pondweed (*Elodea*). Alternatively, a more precise method is to use a photosynthometer which measures the volume of gas produced.

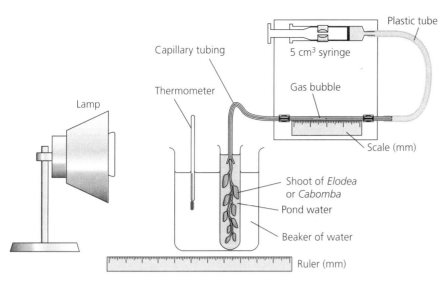

Figure 5.6 A photosynthometer

The apparatus should be set up as shown in Figure 5.6 using aerated water that contains extra sodium hydrogen carbonate as a source of carbon dioxide for the plant. Only one factor should be investigated at a time, keeping the other limiting factors controlled. After a period of equilibration, the following factors can be investigated:

- changing carbon dioxide concentration by adding more sodium hydrogen carbonate to the water
- changing light intensity by moving the lamp closer to or further from the plant
- changing temperature by placing the apparatus in a water bath

> **Examiner's tip**
>
> This topic lends itself to testing How Science Works — interpreting and evaluating the methods and results of investigations.

Tested ☐

Now test yourself

5 When using a photosynthometer, explain why:
   (a) the water used should be aerated
   (b) the water should contain extra sodium hydrogen carbonate
   (c) the apparatus should be left for 5 minutes before readings are made
   (d) using a photosynthometer is more precise than counting bubbles

Answers on p. 123

Exam practice

1 **(a)** Define the term *photosynthetic pigment*. [1]
   (b) Identify the primary photosynthetic pigment and three accessory pigments. [4]
   (c) Explain why most leaves contain more than one photosynthetic pigment. [3]
   (d) Explain why many aquatic plants living at greater depths are dark or almost black in colour. [3]
2 **(a) (i)** Explain the term *photophosphorylation*. [2]
       (ii) Describe how ATP is made through non-cyclic photophosphorylation. In your answer you should use appropriate technical terms, spelt correctly. [5]
   (b) Describe why oxygen is released as a by-product of the light-dependent stage of photosynthesis. [3]
3 **(a)** Identify two products of the light-dependent stage of photosynthesis that are used in the Calvin cycle. [2]
   (b) Describe how carbon dioxide is fixed. [2]
   (c) Explain the effects of low carbon dioxide concentration on the rate of photosynthesis. [3]

4 A student investigated the effect of different coloured light on the rate of photosynthesis. He placed different coloured filters between a light and a piece of pondweed (*Elodea*) and counted the number of bubbles released from the pondweed in 2 minutes. His results are shown in the following table.

Colour of filter	Number of bubbles counted in 2 minutes
Black	12
Blue	78
Green	18
Yellow	32
Red	96

(a) (i) State which colour of light enabled the most rapid rate of photosynthesis. [1]

(ii) Explain why there was very little oxygen released with the black filter. [2]

(iii) Explain why there was little oxygen released with the green filter. [2]

(b) The student expected there to be no oxygen released with the black filter in place. Suggest two possible reasons why some bubbles were observed using the black filter. [2]

(c) Suggest two ways in which the student could have improved his investigation. [2]

Answers and quick quiz 5 online

Online

Examiner's summary

By the end of this chapter you should be able to:

✔ Define the terms *autotroph* and *heterotroph*.

✔ State that light energy is used to produce complex organic molecules in photosynthesis.

✔ Explain how respiration in plants and animals relies on photosynthesis.

✔ State that photosynthesis has two stages and explain how the structure of chloroplasts enables them to carry out photosynthesis.

✔ Define the term *photosynthetic pigment* and explain their importance in photosynthesis.

✔ State that the light-dependent stage takes place in thylakoid membranes and that the light-independent stage takes place in the stroma.

✔ Outline how light energy is converted to chemical energy (ATP and reduced NADP) in the light-dependent stage.

✔ Explain the role of water in the light-dependent stage.

✔ Outline how the products of the light-dependent stage are used in the light-independent stage.

✔ Explain the role of carbon dioxide in the light-independent stage (Calvin cycle).

✔ State that TP can be used to make carbohydrates, lipids and amino acids but that most TP is recycled to RuBP.

✔ Describe the effect on the rate of photosynthesis and on levels of GP, RuBP and TP of changing carbon dioxide concentration, light intensity and temperature.

✔ Discuss limiting factors in photosynthesis.

✔ Describe how to investigate experimentally the factors that affect the rate of photosynthesis.

6 Respiration

Transfer of energy to ATP

The need for respiration
Revised

Respiration is a series of chemical reactions that take place inside all living cells. The reactions release energy from complex organic molecules in a controlled way so that the energy can be used by the cell.

Many processes inside cells require energy and its immediate source is **ATP**. The reactions of respiration produce ATP so that it can be used to drive processes such as **active transport** and energy-requiring **metabolic reactions** such as building large organic molecules.

> **Respiration** is a series of reactions that release energy from substrate molecules inside cells.

ATP
Revised

ATP is a complex molecule containing three components:
- adenine (an organic base)
- ribose sugar (a pentose sugar)
- three phosphate groups

One of the phosphate groups can be removed to leave ADP. The molecule of ADP contains less energy than ATP, so when the phosphate group is removed energy is released. This is the energy used to drive processes in the cell.

Coenzymes
Revised

Coenzymes are complex organic molecules that contribute to enzyme-controlled reactions inside cells. The main role of a coenzyme is to transport the products from one reaction to another reaction.

Respiration is a complex series of reactions and coenzymes are used in a number of places to link these reactions. The coenzymes used include:

> **Coenzymes** are complex organic molecules that are used to transfer the products from one reaction to become the substrate in another reaction.

- **NAD**, which is reduced by hydrogen atoms released from the **respiratory substrate** (usually glucose). The hydrogen atoms can be used in different ways:
  - When oxygen is available, it is transported to the inner membrane of the **mitochondrial cristae**, where they drive an **electron transport chain** involved in ATP production.
  - When oxygen is not available, they may be used in reduction reactions such as reducing **pyruvate** to lactate in **anaerobic respiration**.
- **coenzyme A**, which binds to a 2-carbon **acetate** group in the **link reaction** and delivers the acetate group into **Krebs cycle**.

Respiratory substrates — Revised

A respiratory substrate is a molecule that can be broken down to release energy in respiration. The following molecules can be used, each of which has a different energy value:

- **carbohydrates** (glucose, starch and glycogen) — starch and glycogen are first converted to glucose before entering the respiration pathways.

- **fats** — fats are energy-rich. Fatty acids from a fat are broken down to 2-carbon fragments (acetate), which enter Krebs cycle via coenzyme A. There will be many acetate molecules entering Krebs cycle from one fat molecule. Fats contain many carbon–hydrogen bonds, which means that many molecules of NAD can be reduced by the addition of hydrogen atoms. Each reduced NAD or **FAD** molecule can be used to produce ATP.

- **proteins** — these are converted to amino acids, which are then deaminated. The remaining residue is an organic acid similar to the intermediates in Krebs cycle. These will enter Krebs cycle at an appropriate place. Therefore, each amino acid may produce different yields of ATP. The overall energy yield from proteins is similar to that from carbohydrates.

Now test yourself

1 Explain why fats have more energy per gram than carbohydrates.

Answer on p. 124

Tested

Aerobic respiration

The stages of aerobic respiration — Revised

Aerobic respiration is the release of energy from substrate molecules using oxygen. It is a complex process that takes place in four stages:

1 glycolysis
2 the link reaction
3 Krebs cycle
4 oxidative phosphorylation

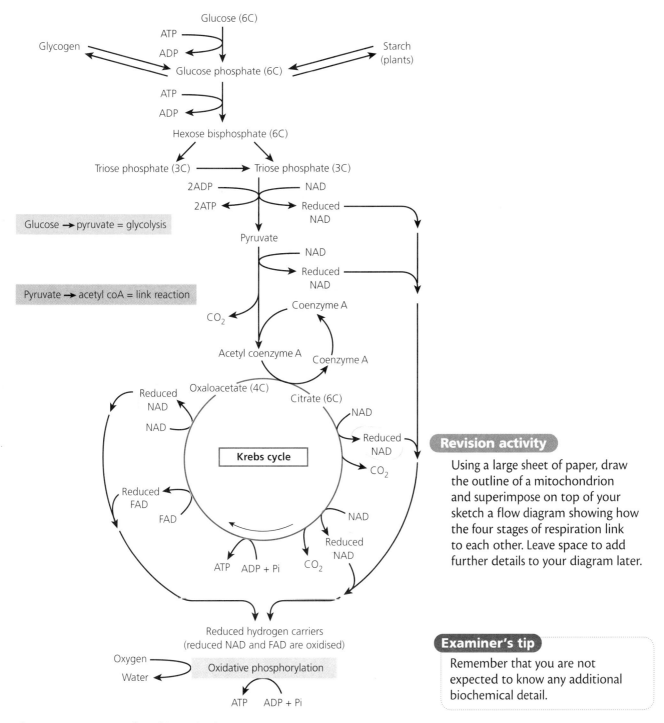

Figure 6.1 A summary of aerobic respiration

Glycolysis

Glycolysis means 'splitting sugar' and it occurs in the **cytoplasm** of every living cell. It is the breakdown of glucose to smaller 3-carbon molecules called pyruvate and does not require oxygen — it is an anaerobic process. The chain of reactions is as follows:

1 **Phosphorylation** of **glucose** by the addition of two phosphate groups from two ATP molecules. This produces **hexose bisphosphate**.

2 The hexose bisphosphate splits into two **triose phosphate molecules**.

Phosphorylation means the addition of a phosphate group.

3 **Oxidation** of the triose phosphate molecules to pyruvate. This is achieved by removal of hydrogen atoms. This enables:

- the reduction of NAD (as the NAD accepts the hydrogen atoms)
- the formation of some ATP by **substrate level phosphorylation**. This is where an enzyme is used to combine ADP and a phosphate group using energy released from the substrate molecule.

The products of glycolysis are:

- reduced NAD
- ATP
- pyruvate

> **Oxidation** means increasing the oxidation number, often by removing hydrogen or electrons.

> **Typical mistake**
> Many candidates forget that glycolysis takes place in the cytoplasm.

> **Revision activity**
> Add this detail to your flow diagram.

The link reaction   Revised ☐

When sufficient oxygen is available for aerobic respiration to take place, the pyruvate is **actively transported** into the **mitochondrion**. The link reaction takes place in the **mitochondrial matrix** and includes a number of steps:

1 **Decarboxylation** of the pyruvate to produce a 2-carbon acetate and carbon dioxide. The carbon dioxide is released for excretion.

2 **Reduction** of **NAD** by hydrogen atoms released from the pyruvate.

3 Combination of the acetate with coenzyme A to produce acetyl coenzyme A, which carries the acetate into the next stage.

> **Decarboxylation** means the removal of a group of atoms including carbon.
>
> **Reduction** means decreasing the oxidation number, often by the addition of a hydrogen atom or electron.

> **Revision activity**
> Add this detail to your flow diagram.

Krebs cycle   Revised ☐

Krebs cycle takes place in the mitochondrial matrix. It consists of a complex series of steps:

1 The acetate is released from the acetyl coenzyme A and combined with a 4-carbon compound called **oxaloacetate**. This forms the 6-carbon compound called **citrate**.

2 The coenzyme A is released to be used again.

3 The citrate enters a cycle of reactions that will release its energy in stages.

4 The citrate is **decarboxylated** and **dehydrogenated** to reform oxaloacetate.

5 Decarboxylation produces carbon dioxide, which is excreted.

6 Dehydrogenation releases hydrogen atoms that are used to reduce the coenzymes NAD and FAD.

7 These reduced coenzymes act as hydrogen carriers to pass the hydrogen on to the next stage.

8 During Krebs cycle, some ATP is produced by substrate level phosphorylation. This is where an enzyme is used to combine ADP and a phosphate group using energy released from the substrate molecule.

> **Dehydrogenation** means to remove a hydrogen atom.

> **Revision activity**
> Add this detail to your flow diagram.

> **Examiner's tip**
> This topic seems very complicated and can be made to sound more complicated by using lots of extra names. However, the only names you need to know in Krebs cycle are oxaloacetate and citrate. If you view the process as a flow diagram, it is easy to follow.

> **Now test yourself**
>
> 2 Explain why it is important that the coenzyme A is released.
>
> Answer on p. 124
>
> Tested ☐

Oxidative phosphorylation

Revised

Oxidative phosphorylation means the addition of a phosphate group in a process that uses oxidation/reduction reactions (Figure 6.2). The inner membrane of the mitochondrion is folded into cristae. The reduced coenzymes carry the hydrogen atoms to the cristae. Embedded in the phospholipid bilayer of the cristae is a series of protein complexes called **electron carriers**. This series of carriers is called an electron transport chain. The first carrier is an enzyme that oxidises the reduced coenzyme by removing the hydrogen and splitting it to a proton (H^+) and an electron (e^-). The coenzyme is released to return into the mitochondrial matrix and accept more hydrogen atoms. The proton enters the mitochondrial matrix. The electron passes from electron carrier to electron carrier down the carrier chain. As an electron passes between carriers, energy is released.

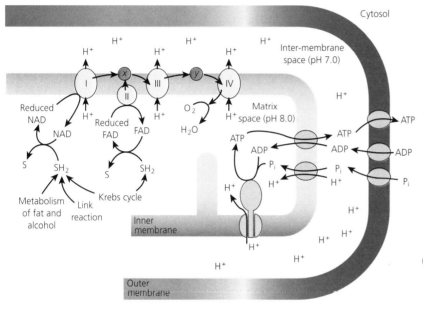

SH_2 = intermediate substances in the link reaction, Krebs cycle and metabolism of fat and alcohol. These substances are dehydrogenated to form reduced hydrogen carriers (reduced FAD and reduced NAD), which are oxidised by complexes I and II.

I, II, III and IV are protein complexes.
x and *y* are electron carrier molecules.

Complexes I, III and IV pump protons; complex II passes electrons from reduced FAD to complex III via electron carrier *x*. Complex II is not a proton pump.

Figure 6.2 Oxidative phosphorylation

Typical mistake

Again, some candidates try to learn too many names — you do not need to know all the names of the carriers.

Revision activity

Add this detail to your flow diagram.

Chemiosmosis

Revised

Chemiosmosis is the process involving the creation of a **proton gradient** through the action of an electron transport chain and the subsequent use of the proton motive force to produce ATP.

The energy released from the electron as it travels down the electron transport chain is used to pump protons from the mitochondrial matrix into the gap between the two mitochondrial membranes. Protons accumulate in the gap to produce a concentration gradient between the gap and the mitochondrial matrix. This is called a proton gradient (it produces a pH gradient and a potential difference across the membrane).

The proton gradient and the potential difference create a driving force called the proton motive force, which pushes the protons out into the mitochondrial matrix. The protons can only pass through special channel proteins attached to the enzyme ATP synthase. As they flow through the ATP synthase, their movement energy is used to produce ATP from ADP and inorganic phosphate. The protons return to the mitochondrial matrix and are combined with electrons from the electron transport chain and oxygen to form water. Therefore, oxygen is the final **electron acceptor** in aerobic respiration.

Revision activity

Add this detail to your flow diagram.

The **experimental evidence** for chemiosmosis includes these points:

- The pH in the gap between the membranes of a mitochondrion is lower (more acidic) than in the inner matrix.
- In chloroplasts, the pH is lower inside the thylakoid spaces.
- Isolated chloroplasts in an illuminated solution of sucrose can turn the solution alkaline (hydrogen ions are pumped into the thylakoid spaces).
- The potential difference across the inner membrane of a mitochondrion is 200 mV (more positive between the membranes compared to the matrix).
- ATP is not made if the stalked particles (ATP synthase) are removed from the cristae.
- ATP cannot be made in mitochondria stripped of their outer membrane.
- ATP is not made in the presence of oligomycin, which blocks the flow of protons through the ion channels.

Now test yourself

3 Explain why oxidative phosphorylation and chemiosmosis cannot occur without oxygen.

4 Explain why a mitochondrion stripped of its outer membrane cannot make ATP.

Answer on p. 124

Tested

Mitochondrial structure

Revised

The mitochondria are highly adapted to enable them to carry out their role in respiration. These adaptations include:

- two membranes separated by an inter-membrane space
- an inner membrane folded into cristae to create a larger surface area
- ATP synthase molecules seen as stalked particles on the cristae
- the presence of electron carrier molecules in the inner membrane
- the presence of the enzymes needed for Krebs cycle in the matrix

Revision activity

Draw a diagram of a mitochondrion and annotate it with notes about how its features enable it to carry out its role in respiration.

ATP yield

Revised

The **theoretical maximum yield** of ATP molecules from glucose is not clear — different texts quote different figures. However, a yield of 32 molecules can be accounted for by assuming that each reduced NAD molecule provides enough energy to make 2.5 ATP molecules and each reduced FAD molecule provides enough energy to make 1.5 ATP molecules.

Examiner's tip

You do not need to account for the number of ATP molecules made.

This theoretical maximum yield is never actually achieved. This is because:

- intermediate substances may be diverted to another reaction pathway
- fats and proteins are also metabolised and acetate is a common intermediate

- energy is used to drive the movement of pyruvate into the mitochondrial matrix
- protons leak out through the outer mitochondrial membrane

Anaerobic respiration

The process of anaerobic respiration

Revised

Anaerobic respiration is the release of energy from substrate molecules without using oxygen and takes place in the cytoplasm. If oxygen is not available, pyruvate does not enter the mitochondrion and the link reaction, Krebs cycle and oxidative phosphorylation do not occur. Therefore, the ATP yield is much lower. The ATP yield from anaerobic respiration is just two molecules from each glucose molecule.

In **mammals**, the pyruvate made in glycolysis is reduced to lactate using the hydrogen from reduced NAD also made in glycolysis. The enzyme lactate dehydrogenase is involved. In **yeast**, the pyruvate is first decarboxylated to ethanal releasing carbon dioxide. The ethanal is then reduced to ethanol by hydrogen from the reduced NAD made in glycolysis (Figure 6.3).

Now test yourself

5   Explain why anaerobic respiration releases less ATP from glucose than aerobic respiration.

Answer on p. 124

Tested

(a) *mammals*

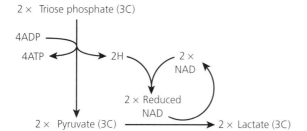

(b) *yeast*

Figure 6.3 The fate of pyruvate in anaerobic respiration: (a) in mammals (b) in yeast

Exam practice

1 **(a)** State precisely where in the cell the following stages of respiration take place.
   (i) glycolysis   [1]
   (ii) Kreb's cycle   [1]
   (iii) oxidative phosphorylation   [1]
   (b) Identify the type of reaction occurring at the following stages of respiration.
   (i) glucose → glucose 6-phosphate   [1]
   (ii) NAD → NADH   [1]

(iii) citrate → 5-carbon intermediate [1]

(iv) pyruvate → lactate [1]

(c) Describe how ATP is produced during oxidative phosphorylation. In your answer you should use appropriate technical terms, spelt correctly. [5]

2 (a) Describe and explain two features of a mitochondrion that enable it to perform its role in respiration. [4]

(b) (i) Explain why anaerobic respiration releases only 2 molecules of ATP compared to 32 molecules released by aerobic respiration. [3]

(ii) Cardiac muscle prefers to respire fatty acids rather than glucose. Suggest why the heart muscle is more affected by limited oxygen supply than other types of muscle. [2]

3 (a) Explain what is meant by the term *chemiosmosis*. [3]

(b) (i) Some protons leak through the outer mitochondrial membrane. Explain what effect such leakage would have on the production of ATP. [3]

(ii) Some of the reduced NAD produced in glycolysis is used to reduce compounds in the cytoplasm. Explain what effect this would have on production of ATP. [3]

4 (a) During oxidative phosphorylation, electrons pass down an electron transport chain in the inner mitochondrial membrane.

(i) State the original source of these electrons. [1]

(ii) State the immediate source of these electrons. [1]

(iii) Name the final electron acceptor. [1]

(b) During electron transfer, electrons pass from one carrier to another, as shown in the following diagram. State which letter (**X** or **Y**) represents oxidation and which represents reduction. [2]

Answers and quick quiz 6 online

Online

Examiner's summary

By the end of this chapter you should be able to:

✔ Outline why all living things need to respire.

✔ Describe the structure of ATP and state that ATP is the immediate source of energy for biological processes.

✔ Explain the importance of coenzymes in respiration.

✔ State that glycolysis takes place in the cytoplasm and outline the process of glycolysis.

✔ State that in aerobic respiration pyruvate is actively transported into the mitochondria and that the link reaction and Krebs cycle take place in the mitochondrial matrix.

✔ Outline the link reaction and Krebs cycle.

✔ Outline oxidative phosphorylation in the cristae.

✔ Outline chemiosmosis and evaluate the experimental evidence for the theory of chemiosmosis.

✔ Explain why the theoretical maximum yield of ATP is not achieved.

✔ Explain why anaerobic respiration produces a much lower yield of ATP than aerobic respiration and compare anaerobic respiration in mammals and yeast.

✔ Define the term *respiratory substrate* and explain the different relative energy values of carbohydrates, proteins and fats.

7 Cellular control

Coding for proteins

The genetic code   Revised

The **genetic code** is the set of rules used to translate information in genetic material to produce proteins. Genes are lengths of **DNA** that code for the structure of **polypeptides**. These polypeptides may be proteins or they may make up part of a protein. These proteins include structural proteins and **enzymes**.

Understanding the genetic code is fundamental to our understanding of how cells and organisms function. The code has a number of features:

- It is a sequence of organic bases in the DNA.
- The bases are read in triplets (groups of three).
- Each triplet codes for a specific **amino acid**, but the code is degenerate, which means that there is more than one code for some amino acids. Some triplets act as stop codes to mark the end of the polypeptide.
- The number of triplets determines the number of amino acids in the polypeptide.
- The sequence of triplets determines the sequence of amino acids.

Typical mistake

Many candidates seem to confuse genes and DNA. Remember that DNA is a large molecule and a gene is a section of that molecule.

Now test yourself   Tested

1   Explain why the bases are read in triplets.

Answers on p. 124

Protein synthesis   Revised

Protein synthesis involves two stages:

1   transcription, which is reading the code and producing a messenger molecule to carry the code out to the cytoplasm
2   translation, which is converting the code to a sequence of amino acids

Transcription

Transcription is the conversion of the genetic code to a sequence of nucleotides in **messenger RNA (mRNA)**. The genetic code is held in the nucleus. The DNA molecule is too large to leave the nucleus, so a smaller messenger molecule is made (mRNA). The double-stranded DNA molecule is unwound and split by the action of the enzyme RNA polymerase, which breaks the hydrogen bonds holding the two strands together. This exposes the bases in the gene. The coding polynucleotide carries the genetic code and the complementary strand is the non-coding strand, sometimes called the antisense strand or the template strand. The template strand is used to build a copy of the coding strand. The copy of RNA is made, which is a single-stranded nucleic acid in which the sugar is ribose (rather than deoxyribose) and the base thymine is replaced by uracil. Apart from these differences, the mRNA molecule is a copy of the coding strand so it has an identical sequence of bases to the coding strand of DNA.

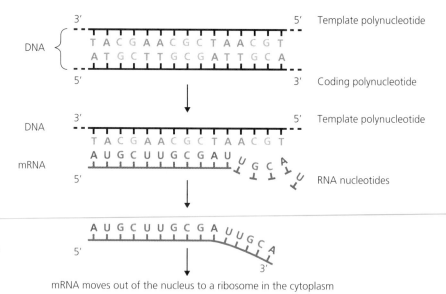

Figure 7.1 Transcription of the template polynucleotide in DNA

Each triplet of bases on the mRNA is called a codon. The enzyme RNA polymerase joins the bases of the RNA to produce a complete single-stranded molecule (Figure 7.1). The molecule of mRNA detaches from the DNA template and the DNA strands can rejoin to make the double helix. The mRNA leaves the nucleus via the nuclear pores and enters the cytoplasm.

Table 7.1 Base pairing

DNA template polynucleotide	mRNA
A	U
T	A
C	G
G	C

Now test yourself

2 Explain why the genetic code must be converted to mRNA before it leaves the nucleus.

Answer on p. 124

Tested

Translation

Translation is the conversion of the code in mRNA to a sequence of amino acids. Amino acids must be activated, so they are combined with a specific molecule of **transfer RNA (tRNA)** that has a specific base triplet called the anticodon. The anticodon of the tRNA is complementary to a codon on the mRNA. ATP is used in this activation process.

The mRNA joins on to a **ribosome** in the cytoplasm. The ribosome contains enough space for two codons at a time. The mRNA slides through the groove between the two components of the ribosome. As each codon enters the ribosome, it is used to position the next amino acid. The amino acids attached to tRNA molecules are aligned with the correct part of the mRNA by the complementary pairing of the bases in the codon and anticodon. Enzymes bind the amino acids together in a chain by condensation reactions to create a growing polypeptide chain.

The tRNA molecule is then released to be reused. When the ribosome reaches a stop codon, the complete polypeptide chain is released and folds to form the secondary and tertiary structure of the protein.

Some proteins need to be activated by **cyclic adenosine monophosphate (cAMP)**, which interacts with the new protein to alter its **three-dimensional structure**. This makes the proteins a better shape to fit their complementary molecules.

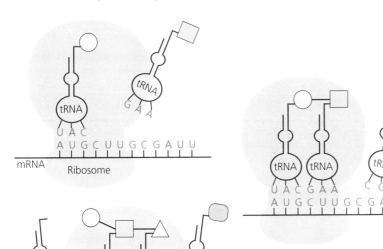

Figure 7.2 Translation

Revision activity

Sketch a diagram of a cell including the organelles relevant to protein synthesis, then add a flow diagram of protein synthesis over the sketch.

Examiner's tip

Protein synthesis is a long sequence of events, so a quality of written communication mark could be given for correct sequencing in your description.

Now test yourself

3 Explain why mRNA must be single-stranded.

Answer on p. 124

Tested

Mutations

Revised

Mutations are changes in the **nucleotide sequence** in the DNA molecule, resulting in a different base sequence. There are a variety of types of mutations, which can each have a different effect.

Substitution

Substitution is where one base pair is changed to a different base pair. This alters the code for that triplet. It is likely that the new triplet may code for a different amino acid, in which case one amino acid in the polypeptide will be different from the normal sequence. However, due to the degenerate nature of the genetic code, some amino acids have more than one triplet as a code. Therefore, the polypeptide may not be affected at all.

Deletion

Deletion is where one or more base pairs are deleted from the sequence. This causes frameshift, as it may alter every single triplet downstream of the change. If one or two base pairs are deleted, every single triplet downstream of the change will be affected. This changes the sequence of amino acids in the polypeptide and the polypeptide or protein produced is very different from its original form. However, if three base pairs (or a multiple of three) are deleted, all the subsequent triplets are the same and the amino acid sequence is not changed as much — the polypeptide simply misses one or more amino acids.

Addition

Addition is where one or more base pairs are added to the sequence. This has a similar effect to deletion except that amino acids may be added to the polypeptide, which therefore gets longer.

Stutter

Stutter is where triplets are repeated, adding extra amino acids to the polypeptide.

Mutation effects
Revised

The effect that a mutation has depends on the type of mutation. Mutations can be **neutral** or **beneficial**, but most are **harmful**. Since most mutated genes are recessive, it is likely that the effects of a mutation will be masked by the production of the normal protein using the dominant allele.

Neutral mutations

Neutral mutations are those that have no effect. This could be because:

- the mutant triplet codes for the same amino acid
- the mutant triplet changes the amino acid but this has no effect on the function of the polypeptide, or its change in function produces no advantage or disadvantage
- the mutation occurs in non-coding parts of the DNA

Beneficial mutations

Occasionally, a mutation may alter the polypeptide such that it works more effectively, but this is unlikely. However, a different version of the gene (an allele) that produces a different version of the polypeptide may become advantageous if the environment changes. This is the basis of variation, allowing natural selection and evolution to occur.

Harmful mutations

Most mutations are harmful. They alter the structure of the polypeptide such that it works less well or does not work at all. It is important to remember that even those mutations that confer a disadvantage at first may become advantageous if the environment changes.

> **Typical mistake**
>
> Many candidates seem to believe that all mutations are harmful or even lethal.

> **Revision activity**
>
> Draw a mind map about mutations. Start with the structure of DNA at the centre.

> **Examiner's tip**
>
> Remember that mutations are rarely lost — they are often recessive and hidden by the dominant allele. This is important for the success of evolution.

> **Now test yourself**
>
> 4 Explain why most mutations are harmful.
>
> Answer on p. 124
>
> Tested

The *lac* operon
Revised

The ***lac* operon** is a functional unit of genes in the genome of **prokaryotic cells** found in the bacterium *E. coli*. It contains two genes that code for the structure of proteins and a **genetic control** mechanism to enable the genes to be switched on and off. Under normal conditions the bacterium is not able to digest the sugar lactose. This is because it does not make the enzyme β-galactosidase. However, it has the ability to make β-galactosidase and the gene is switched on when lactose is available. This is called induction. The *lac* operon consists of three regions:

- two structural genes
- a promoter region
- an operator region

There is also a separate regulator gene elsewhere. This codes for a repressor protein, which binds to the operator region of the *lac* operon. This prevents the binding of the enzyme RNA polymerase, which starts the process of transcription of the structural genes. Therefore, under normal conditions, structural genes are not expressed.

When lactose is available, it binds to the repressor protein and changes its shape. The repressor protein can no longer bind to the operator region, but RNA polymerase can bind to the promoter region and start transcription. The two structural genes are expressed and β-galactosidase is produced along with another protein called lactose permease, which makes the cell membrane more permeable to lactose.

Now test yourself

5 Suggest why it is useful to the bacterium to be able to switch a gene on and off.

Answer on p. 124

Tested ☐

(a) High concentration of glucose, low concentration of lactose

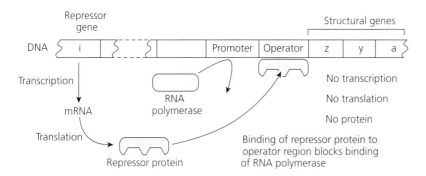

(b) Low concentration of glucose, high concentration of lactose

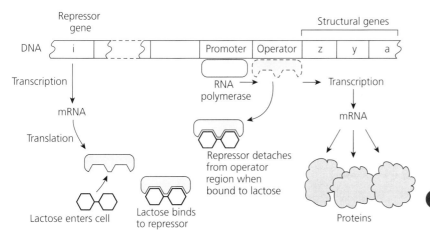

Figure 7.3 How the *lac* operon functions: (a) when lactose is absent and glucose is present in high concentrations (b) when lactose is present and there is little or no glucose

Typical mistake

Some candidates fail to learn this detail and simply suggest that lactose activates the structural genes.

Genes that control development

Homeobox sequences
Revised

Homeobox genes control the basic structure and orientation (**body plan**) of an organism. This is achieved by controlling the differentiation of cells and parts of the body through switching genes on and off at appropriate times during development. As the homeobox genes are activated, they activate structural genes in a carefully coordinated sequence to ensure that the features develop in the correct way.

> **Homeobox genes** control the body plan of an organism.

The homeobox genes are activated in a particular order that matches the order in which they are expressed along the body from head to tail. They are arranged into groups called hox clusters. The fruit fly *Drosophila* has two clusters. Cluster A controls the development of the head and thorax whereas cluster B controls development of the thorax and abdomen. Humans have four clusters, each with 9 to 11 genes. The number of clusters and the number of genes in each cluster reflect the complexity of the body plan.

A homeobox gene contains a sequence of 180 base pairs known as a **homeobox sequence**. This sequence codes for a sequence of 60 amino acids, which is found in the polypeptide produced. The polypeptides produced are transcription factors that bind to genes and initiate their transcription. Homeobox genes are similar in all **plants**, **animals** and **fungi**. This is because they have the same role in each case: they code for transcription factors that need to bind to DNA.

> **Typical mistake**
> Some candidates do not distinguish between homeobox genes and homeobox sequence.

Apoptosis
Revised

Apoptosis (**programmed cell death**) is a series of carefully controlled biochemical events that leads to orderly cell death. The sequence of events is as follows:

1 Enzymes break down the cytoskeleton.

2 The cell shrinks and organelles are packed together, along with fragments of the chromatin.

3 The cell surface membrane breaks up to form cell fragments (vesicles) containing the cell contents.

4 The vesicles are taken up by phagocytes and digested.

Apoptosis is an important part of development, as tissues may need to be modified during development. During growth of a multicellular organism, cells may need to be killed or removed to ensure correct development. For example:

● T lymphocytes that recognise our own body antigens would attack our own cells if they were allowed to survive and become active.

● During development of the hands and feet, tissue grows between the fingers and toes. This must be removed at a later stage to separate the fingers and toes.

Exam practice answers and quick quizzes at **www.therevisionbutton.co.uk/myrevisionnotes**

Exam practice

1 **(a)** Describe three ways in which the structures of RNA and DNA are different. [3]

(b) (i) Explain why the genetic code is said to be a triplet code. [2]

(ii) Explain how some mutations may have no effect on the organism. [4]

(c) Describe how polypeptides are synthesised and explain how the structure of a polypeptide is determined by the sequence of nucleotide bases in the gene. In your answer you should make clear the sequence of events in protein synthesis. [10]

2 **(a)** The original base sequence in a gene and two mutated versions of the base sequence are shown below:

- original sequence: AGTTTCGCCCGT
- mutated sequence 1: AGTCTCGCCCGT
- mutated sequence 2: AGTTCGCCCGT

(i) Name the type of mutation that has occurred in mutated sequence 1. [1]

(ii) Suggest what effect this mutation may have on the functionality of the polypeptide produced. [2]

(iii) Name the type of mutation that has occurred in mutated sequence 2. [1]

(iv) Suggest what effect this mutation may have on the functionality of the polypeptide produced. [2]

(b) A mutation such as that seen in mutated sequence 2 could occur in the homeobox sequence of a homeobox gene. Explain why this would have far-reaching effects on the organism. [6]

3 **(a)** State what is meant by the term *apoptosis*. [1]

(b) Describe the process of apoptosis. [4]

(c) Explain how apoptosis can contribute to the development of a body plan. [3]

Answers and quick quiz 7 online

Online

Examiner's summary

By the end of this chapter you should be able to:
- State that genes code for polypeptides, including enzymes.
- Explain the meaning of the term *genetic code* and describe how a nucleotide sequence codes for the amino acid sequence in a polypeptide.
- Describe how the sequence of nucleotides within a gene is used to construct a polypeptide.
- State that mutations cause changes to the nucleotide sequence and can be beneficial, neutral or harmful.
- State that cAMP activates proteins.
- Explain the genetic control of protein production in the *lac* operon.
- Explain that the genes controlling the development of body plans are similar in plants, animals and fungi.
- Outline how apoptosis can change body plans.

8 Meiosis and variation

Meiosis

Meiosis is the division of the nucleus to produce four haploid nuclei.. Before division starts (during interphase), the DNA replicates so that each **chromosome** consists of two identical copies called sister chromatids that are held together by a centromere. The cell divides twice to produce four daughter cells, each of which contains half the number of chromosomes as the parent cells.

Stages of meiosis

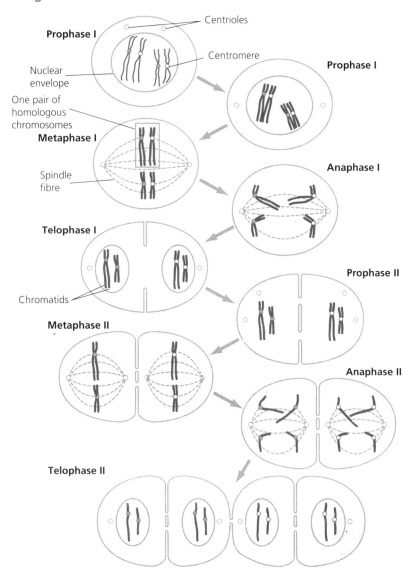

Figure 8.1 Meiosis to form sperm cells in an animal that has a diploid number of 4. Maternal chromosomes are blue, paternal chromosomes are red

Examiner's tip

This examination paper includes synoptic marks. These test your:
- understanding of the principles behind different processes
- ability to make links back to other parts of the specification

The obvious links here are:
- variation, causes of variation, selection and evolution from F212

Typical mistake

Many candidates confuse homologous chromosomes with sister chromatids. Remember that homologous chromosomes are the two chromosomes of similar length found in all diploid cells. They may contain identical alleles or alleles that are slightly different. Sister chromatids are the two identical copies of the same chromosome that are produced by replication of the DNA before cell division starts.

Exam practice answers and quick quizzes at **www.therevisionbutton.co.uk/myrevisionnotes**

Prophase I — chromosomes coil tightly to become shorter and thicker in a process known as condensing. Homologous chromosomes pair to form **bivalents**, each containing four chromatids. The chromatids in each bivalent break and rejoin to form **chiasmata** or crossovers — this is where sections of the non-sister chromatids can be exchanged. The **nuclear envelope** breaks up. The centrioles migrate to opposite poles of the cell to form the spindle.

Metaphase I — microtubules (spindle fibres) attach from the centrioles to the centromere of each chromosome. The bivalents move to the equator of the cell. Orientation of each bivalent on the equator is random — maternal or paternal chromosomes could be facing either pole.

Anaphase I — the microtubules shorten to separate the homologous chromosomes and pull them towards opposite poles. Each chromosome still consists of two chromatids.

Telophase I — the chromosomes reach opposite poles. The nuclear envelope reforms around each set of chromosomes to produce two nuclei. These nuclei are haploid as they have one chromosome from each homologous pair (but there are still two sister chromatids). The **cell membrane** pinches in to form two cells — this is cytokinesis.

Prophase II — the nuclear membranes break up again. The centrioles replicate again and migrate to opposite poles of the two new cells.

Metaphase II — the microtubules attach between the centrioles and the centromere of each chromosome. The chromosomes move to the equator and align randomly.

Anaphase II — sister chromatids move to opposite poles.

Telophase II — the nuclear membranes reform. Cytokinesis occurs to produce four genetically different haploid cells.

A **bivalent** is one pair of homologous chromosomes.

Chiasmata (singular: chiasma) are the points where two non-sister chromatids exchange genetic material during crossover.

> **Examiner's tip**
>
> Make sure you can spell the terms accurately, and in particular *meiosis*, *centriole* and *centromere*.

> **Examiner's tip**
>
> Many candidates are unsure about the terms *chromosome* and *chromatid*. A chromosome replicates to form two identical chromatids. While the two copies are together during prophase and metaphase, we refer to the structure consisting of two chromatids as a chromosome. However, the two chromatids separate during anaphase and once they are apart each one is called a chromosome.

> **Revision activity**
>
> Draw your own set of diagrams showing meiosis and annotate each diagram to show what is occurring at each stage.

Now test yourself   Tested ☐

1   Explain why there must be two divisions in meiosis.

Answers on p. 124

Definitions of key terms ——————————————— Revised ☐

Allele — an alternative form of a gene, found at the same position on homologous chromosomes. In most cases an organism has two copies of every gene, which could be identical copies or alleles. For example, there might be a gene for coat colour in rats, with a *B* allele for brown fur and a *b* allele for albino fur.

Codominant — a characteristic in which both alleles contribute to the phenotype.

Crossing over — where chromatids break and rejoin during prophase I of meiosis. If crossing over occurs between non-sister chromatids, this means alleles on one chromatid are exchanged with alleles on the other chromatid. This results in the recombination of alleles and produces genetic variation.

Dominant — a characteristic that masks another characteristic. The allele responsible is expressed in the phenotype through production of a protein.

Genotype — the combination of alleles held in the nucleus.

Linkage — where different genes for different characteristics are linked together so that the characteristics often appear together in the phenotype. The genes are found close together on the same chromosome so that they usually stay together during meiosis. Only rarely does a cross-over occur between these two genes to produce a recombinant.

Locus — the position of a gene on the chromosome.

Phenotype — the characteristics expressed in the individual. These characteristics result from the proteins produced by the genes.

Recessive — a characteristic that is masked by another characteristic and is only expressed in the phenotype if no dominant allele is present.

Now test yourself

Tested ☐

2 Explain the difference between the following pairs of terms:
   (a) dominant and codominant
   (b) chromosome and chromatid
   (c) gene and allele

Answers on p. 124

Creating variation

Revised ☐

Genetic **variation** can be created by meiosis. When chromatids cross over, they exchange lengths of DNA. When this occurs between non-sister chromatids, it produces new combinations of alleles (Figure 8.2).

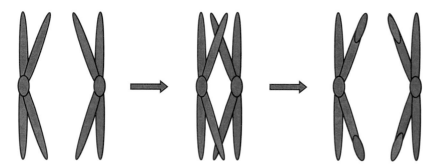

Figure 8.2 The exchange of genetic material in non-sister chromatids

The way the bivalents orientate on the equator during metaphase I is random. This means that either the maternal or the paternal chromosome of a bivalent may face either pole. Therefore, the combination of maternal and paternal chromosomes migrating to either pole is random. This is called **independent assortment** of homologous chromosomes (Figure 8.3). In a similar way, the orientation of the chromosomes on the equator in metaphase II is random. Therefore, the combination of chromatids migrating to each pole is random. This is called independent assortment of sister chromatids.

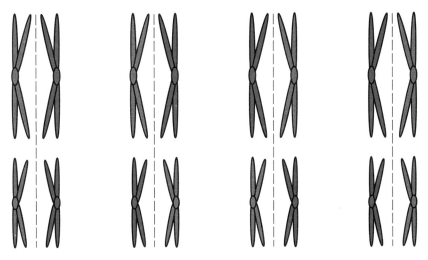

Figure 8.3 Four possible combinations arise due to independent assortment of homologous chromosomes on the equator

Examiner's tip

In a question about creating variation, remember that random assortment occurs twice: once in metaphase I and again in metaphase II.

During replication of DNA to form chromatids, the copying of the nucleotide sequence may be inaccurate. This is a mutation. If it occurs in the cells about to undergo meiosis, it produces variation.

Fertilisation is the combination of the paternal DNA in a sperm with the maternal DNA in the egg. This is essential to re-establish the diploid number of chromosomes. As a result of meiosis, the combination of alleles in every egg cell is slightly different. Equally, the combination of alleles in every sperm cell is different. Any sperm cell can fertilise any egg (random fertilisation), so the potential for increased genetic variation is huge.

Typical mistake

Many candidates treat fertilisation as if it were part of meiosis, but this is not the case. Meiosis is the division of the nucleus and formation of haploid cells. Fertilisation is the combination of two haploid cells to produce a single diploid cell.

Genetic diagrams

Using genetic diagrams

Revised

Genetic diagrams are a clear way to set out and explain the results of a cross. There are certain rules you should follow:

- Always use the standard format for diagrams.
- Start with the full phenotype and genotype of the parents.
- Remember that each gamete contains one gene from each pair of alleles in the parent.
- Draw a circle around the genes in a gamete to signify it is a gamete.
- Remember to combine each male gamete with each female gamete — this represents random fertilisation. Using a Punnett square will help.
- Write out the full phenotypes of the offspring based on the genotypes produced in the diagram.

Examiner's tip

When drawing a genetic diagram, make sure each step is clear. Follow the standard format described here and use the same symbols for genes given in the question.

Sex linkage

Revised

Sex linkage is seen where certain genes are found on only one of the chromosomes that determine sex. This usually means that the Y chromosome is missing a gene and a male (XY) has only one copy of the gene.

Sex linkage is where a certain characteristic occurs more often in one gender than the other. It is caused by the presence of the gene on only one sex chromosome.

In diagrams involving sex linkage, the X and Y chromosomes are included with the relevant genes shown as superscript. Figure 8.4 shows the inheritance pattern found when a homozygous white-eyed female fruit fly (*Drosophila melanogaster*) is crossed with a red-eyed male. Red eye is the dominant feature (allele R), whereas white eye is recessive (r). The offspring (F_1) of the original parents are then crossed to produce an F_2 generation.

Parental phenotypes	Red-eyed male	$\times$	White-eyed female
Parental genotypes	$X^R Y$		$X^r X^r$
Parental gametes	X^R Y		X^r

	Female gametes	
		X^r
Male gametes	X^R	$X^R X^r$
	Y	$X^r Y$

F_1 genotype(s)	$X^r Y$	$X^R X^r$
F_1 phenotype(s)	White-eyed male	Red-eyed female

F_1 genotypes	$X^r Y$	$\times$	$X^R X^r$
F_1 gametes	X^r Y		X^R X^r

	Female gametes	
	X^R	X^r
Male gametes X^r	$X^R X^r$	$X^r X^r$
Y	$X^R Y$	$X^r Y$

F_2 genotypes	$X^R Y$	$X^r Y$	$X^R X^r$	$X^r X^r$
F_2 phenotypes	Red-eyed male	White-eyed male	Red-eyed female	White-eyed female
F_2 ratio	1 :	1 :	1 :	1
	red-eyed male	white-eyed male	red-eyed female	white-eyed female

Figure 8.4 A genetic diagram to show sex linkage in *Drosophila melanogaster*

Now test yourself

Tested ☐

3 Draw a genetic diagram to show the expected phenotype ratio in the F_2 generation when a pure breeding (homozygous) red-eyed female fly is crossed with a white-eyed male and the F_1 are crossed.

Answers on p. 124

Codominance

Revised ☐

Codominance is seen where there is no dominant and no recessive characteristic, so both alleles contribute to the phenotype. For example, in snapdragons flower colour can be red (genotype RR), white (genotype WW) or pink (genotype RW). Both alleles R and W are equally dominant (Figure 8.5).

> **Codominance** is seen where both alleles contribute to the phenotype.

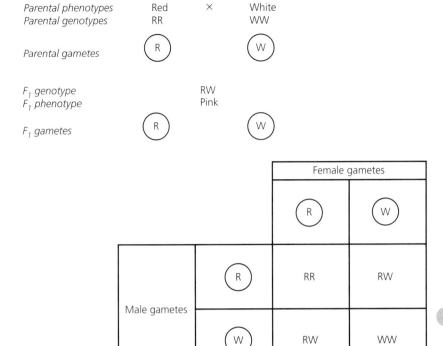

| Parental phenotypes | Red | × | White |
| Parental genotypes | RR | | WW |

| F_1 genotype | RW |
| F_1 phenotype | Pink |

		Female gametes	
		R	W
Male gametes	R	RR	RW
	W	RW	WW

F_2 genotypes	RR	WW	RW
F_2 phenotypes	Red	White	Pink
F_2 ratio	1 red	: 1 white	: 2 pink

Figure 8.5 A genetic diagram to show codominance in snapdragons

Now test yourself

4 Draw a genetic diagram to show why you would not expect to get any white offspring when a roan cow is crossed with a red bull. (Coat colour is a codominant feature.)

Answer on p. 125

Tested

Interactions between loci

Epistasis   Revised

Epistasis is when two or more genes (on different **loci**) influence the phenotype. This is seen where a chain of reactions (a reaction pathway) leads to production of the protein that produces the phenotype. Each step in the pathway requires an enzyme and each enzyme is produced by a different gene.

> **Epistasis** is where genes on different loci interact.

Recessive epistasis

To make a purple compound, both enzymes A and B are needed. Therefore, the organism must possess dominant alleles of both genes A and B (see Figure 8.6).

If allele A is not present, enzyme A is not made and the precursor is not converted to a red compound. If allele A is present but allele B is not present, enzyme A is made but enzyme B is not made. The red compound cannot be converted to a purple compound. If allele A is not present, it does not matter if allele B is present or not as there will be no red compound to be converted to purple compound. This is called recessive epistasis — the presence of recessive alleles for one gene (gene A) affects the expression of another gene (gene B).

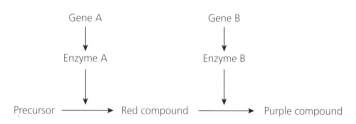

Figure 8.6 Recessive epistasis

Dominant epistasis

To make a red compound, enzyme A is needed from gene A (the dominant allele A must be present). However, if the dominant allele of gene B is present, it produces enzyme B. Enzyme B competes with enzyme A for the substrate. If it is a better competitor, more of the precursor is converted to purple rather than to red. This is called dominant epistasis because the presence of the dominant allele of a gene (gene B) affects the expression of another gene (gene A). Obviously, if only recessive alleles of gene B were present (bb), gene A would not be affected.

Predicting ratios

Predicting **phenotypic ratios** is most easily achieved by using a simple diagram or a Punnett square.

- Write down the genotypes.
- Work out the gametes from the parental genotypes.
- Insert the gametes in a Punnett square.
- Work out the genotypes of the offspring.
- Ensure you take account of the epistatic effects in interpreting the phenotype:
  - In recessive epistasis, there must be a dominant allele of one gene in order for the second gene to be expressed.
  - In dominant epistasis, there must be no dominant alleles of one gene in order for the second gene to be expressed.

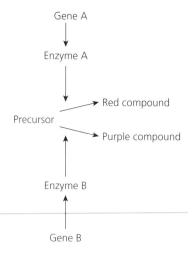

Figure 8.7 Dominant epistasis

Typical mistake

Many candidates seem to confuse dominant and recessive epistasis. Unless you are asked to state whether it is dominant or recessive, simply call any gene interaction epistasis without giving further detail.

Examiner's tip

You are not expected to draw genetic diagrams for examples of epistasis. However, you may be asked to interpret genetic diagrams of epistasis.

The chi-squared test

Revised

The **chi-squared test** is a statistical test used to determine whether the observed results of a cross (O) differ significantly from the expected results (E). The expected results are theoretical and usually calculated using a genetic diagram. The formula is:

$$\chi^2 = \sum \frac{(O-E)^2}{E}$$

- This formula calculates a value for the difference between O and E, which can then be used to determine whether the difference is due to chance.
- The larger the value of χ^2 calculated, the greater the difference between the observed and expected results and therefore the lower the probability that the observed results were created by chance.
- The way the test is used involves making a null hypothesis. This null hypothesis may be 'there is no difference between the observed and expected results'.

Examiner's tip

You are unlikely to be asked to carry out a complete calculation in an examination. You may need to carry out part of a calculation, but it is more likely that you may be asked to discuss the use of the test or to interpret the results of a test.

- Once the value of χ^2 is calculated using the formula, it is compared to a critical value of χ^2 which can be looked up on a table of probabilities for chi-squared.
- If the calculated value of χ^2 is lower than the critical value, we can accept the null hypothesis as the differences could arise due to chance. However, if the calculated value of χ^2 is greater than the critical value, there is a low probability that the differences could arise by chance. The observed results are significantly different from the expected.

Variation

Continuous and discontinuous variation
Revised

Continuous variation is seen where there are no distinct groups or categories and there is a full range between two extremes, for instance in height and body weight. The continuously variable feature can be quantified and data are usually presented in the form of a histogram. This form of variation is caused by:

- a number of genes interacting together — the combination of at least two alleles for each gene and a number of genes all affecting one characteristic means that there are likely to be a lot of different variations in the final phenotype expressed
- the environment

Continuous variation is variation that shows a complete range with no distinct groups.

Discontinuous variation is seen where there are distinct groups or categories and there are no in-between types, for instance in gender and possession of resistance or immunity. The discontinuously variable feature cannot be quantified — it is qualitative. Data are usually presented in the form of a bar chart.

This type of variation is usually caused by the presence of a dominant allele producing one phenotype, whereas the absence of that dominant allele in the double recessive is expressed as a different phenotype.

Discontinuous variation is variation that produces distinct groups.

Typical mistake

Candidates tend to plot bar charts with no spaces between the bars, but there should be gaps between them.

Causes of variation
Revised

There are two causes of **phenotypic variation**:

- genetic
- environmental

Many variable features may be affected by both causes. For example, skin colour in humans is genetically determined. However, exposure to the sun will cause extra pigment production producing a tan.

Genetic causes of variation are a result of differences in the sequence of bases in the DNA. They are caused by mutations that arise spontaneously and randomly, and are passed on from one generation to the next. They usually cause discontinuous variation. Examples include:

- number of limbs
- eye colour
- ability to roll the tongue

Environmental causes of variation are caused by variations in exposure to certain environmental conditions. They are not passed on from one generation to the next and cause continuous variation. Examples include:

- skin colour caused by exposure to sunlight
- body mass

The need for variation

Variation is an essential part of evolution, which is achieved by natural selection. This is the selection of the best-adapted individuals, allowing them to survive and pass on their genes to the next generation. If there is no variation between individuals, none will be better adapted than any others and selection cannot occur.

Revision activity

Draw a mind map with 'variation' at the centre. Leave space to continue expanding the diagram later.

The Hardy-Weinberg principle

Revised

The **Hardy-Weinberg principle** is used to calculate **allele frequencies** in a population. A population contains individuals of three genotypes: AA, Aa and aa, where A is the dominant allele and a is the recessive allele. The formulae involved are as follows:

Formula 1: $p + q = 1$

Formula 2: $p^2 + 2pq + q^2 = 1$

where p is the frequency of the dominant allele and q is the frequency of the recessive allele. Therefore, p^2 is the frequency of homozygous dominant individuals (AA), $2pq$ is the frequency of heterozygous individuals (Aa) and q^2 is the frequency of homozygous recessive individuals (aa).

The individuals with the homozygous recessive genotype can be recognised — they show the recessive phenotype. In a population where 9% of individuals show the recessive feature, this is a frequency of 0.09. Therefore, $q^2 = 0.09$ and $q = \sqrt{0.09} = 0.3$.

To find the value of p, we enter $q = 0.3$ into formula 1: $p + 0.3 = 1$. Therefore, $p = 0.7$.

Knowing the values of p and q, we can calculate the frequencies of the different genotypes in the population:

The frequency of the homozygous dominant (p^2) is 0.7^2

$= 0.49$

The frequency of the heterozygous genotype ($2pq$) is $2 \times 0.7 \times 0.3$

$= 0.42$

This means that the population consists of 9% homozygous recessive individuals, 49% homozygous dominant individuals and 42% heterozygous individuals.

The Hardy-Weinberg principle assumes that:

- the population is large
- mating is random

Now test yourself

5   Calculate the frequency of heterozygous individuals in a population where 84% of the population show the dominant characteristic.

Answer on p. 125

Tested

- there are no mutations, immigration or emigration
- no **selection pressure** is in operation

> **Selection pressure** is a force that gives a selective advantage to one phenotype over another.

Selection and genetic drift

Selective forces
Revised

A selective force is some aspect of the environment that places a selection pressure on the species, making it more difficult for individuals to survive and reproduce. Therefore, those individuals with better adaptations survive more easily and are more likely to reproduce. This allows them to pass on their alleles to the next generation. Those who are less well adapted struggle to survive and are less likely to pass on their alleles.

We can view all the alleles in a population as a gene pool. Each individual contributes two alleles to the gene pool. If many individuals possess one particular allele, that allele makes up a high proportion of the gene pool — it has a high frequency. If the frequency of alleles in the gene pool remains constant, the population does not change. However, if the frequency of alleles changes, the population changes and evolution is occurring.

Stabilising forces

Assuming that a species is well adapted to the environment and the environment stays constant, there will be no evolutionary change. This is because any change away from the well-adapted form is unlikely to be successful and will not pass on more alleles to the next generation. The alleles that produce these extremes are not likely to increase their proportional share of the gene pool — they do not change in frequency.

One example of stabilising selection is the higher mortality of human babies at extreme birth weights. Both high and low birth weights have a higher mortality, so mean birth weight does not change.

Evolutionary forces

Evolution occurs when the environment changes and places a selection pressure on the species. Genetic variation is caused by the presence of different alleles in the population. If the selection pressure favours those individuals who possess a particular allele, those individuals are placed at a selective advantage. They will reproduce more frequently and a higher proportion of the next generation will possess that allele. The frequency of alleles changes. This is known as evolutionary or directional selection.

> **Examiner's tip**
> Always refer to changes in allele frequency rather than gene frequency.

One example of evolutionary selection is the change from light to dark colour seen in the peppered moth (*Biston betularia*) during the industrial revolution.

Genetic drift
Revised

As individuals are born or die, there may be small changes in the frequency of each allele. In a large population this has little or no effect.

However, in a small population the loss of one or two individuals could have a large impact on the gene pool — it may even cause the loss of a particular allele. Alternatively, the survival of one or two unusual individuals could increase the frequency of a certain allele. These changes are known as **genetic drift**.

> **Genetic drift** is when allele frequencies change by chance.

Isolating mechanisms

The evolution of a new species occurs when members of the same species are unable to interbreed freely. This is called reproductive isolation. When this occurs, any genetic changes do not spread throughout the species and one population may accumulate changes not seen in another population. If enough changes occur that the members of one population can no longer breed successfully (producing fertile offspring) with the members of another population, speciation has occurred.

There are a number of possible **isolating mechanisms**:

- **Ecological (geographic) mechanisms** such as large rivers, seas or mountains, which prevent the populations mixing.
- **Seasonal (temporal) mechanisms** such as variation in the timing of the breeding season each year, which means the populations are not synchronised.
- **Reproductive mechanisms** such as differences in courtship rituals — these rituals are usually stereotyped and a slight change in behaviour or responses can lead to the ritual failing. Another reproductive barrier could be a slight biochemical change, which means that the egg and sperm are not compatible.

Species concepts

There are different ways to define a species:

- **biological species concept** — a group of organisms that can interbreed to produce fertile offspring
- **phylogenetic (cladistic/evolutionary) species concept** — a group of organisms that share many characteristics (morphology, physiology, embryology and behaviour) and occupy the same ecological niche

The biological species concept is difficult to apply if no studies on reproductive capability have been carried out and with organisms that reproduce asexually. Therefore, a taxonomist trying to classify a fossil or preserved specimen cannot use this definition.

However, there may be significant variation between individuals in a species. For example, males and females may look different; the larval stage of a butterfly (a caterpillar) looks very different from the adult. Therefore, the phylogenetic species concept can also be difficult to apply. The cladistic approach uses many sources of information including modern research involving the sequencing of DNA and certain proteins. This approach applied to the phylogenetic species concept has often proven to be the most reliable way of defining a species.

Natural and artificial selection

Natural selection and **artificial selection** operate in the same way — a selection pressure that causes allele frequencies to change:

- In natural selection, the selective forces are biotic or abiotic aspects of the environment. An allele that confers a selective advantage to the individual will be selected as that individual is more likely to survive and breed to pass on that allele to the next generation. Natural selection affects all the features of an organism and proceeds slowly unless the environment changes dramatically.

- In artificial selection, the selective forces are humans. An individual that shows a desired characteristic is selected to breed. This means that the allele causing that desired characteristic is passed on to the next generation. Artificial selection affects only the one or two characteristics chosen by the human and can proceed quickly as selection pressure is very strong.

Artificial selection of modern dairy cows

Selection can take a long time, as cows take many months or years to reach their full milk-producing potential. Milk yield shows continuous variation. Females are selected according to their performance — a high milk yield is desirable. Males are selected by testing the performance of their female offspring. Once a suitable bull has been identified, its sperm can be collected and stored (possibly being frozen for many years). The sperm can be delivered by artificial insemination to a large number of suitable females. The offspring can then be tested and over a number of generations the milk yield per cow can be increased.

> **Typical mistake**
> Many candidates omit to state that selection must continue over several generations.

Artificial selection of bread wheat

Modern bread wheat (*Triticum aestivum*) is the result of artificial selection over thousands of years:

- Einkorn wheat (*Triticum urartu*) crossed naturally with wild grass (*Aegilops*) to produce a hybrid. This hybrid was sterile, but became fertile after a chromosome mutation. This was wild emmer wheat (*Triticum turgidum dicoccoides*).

- Selection by farmers in the Middle East 10 000 years ago resulted in the wild emmer wheat giving rise to cultivated emmer wheat (*Triticum turgidum dicoccum*).

- This cultivated emmer wheat was bred with other species (*Aegilops tauschii*) to produce a number of hybrids. These hybrids were sterile because they possessed an uneven number of chromosomes, which meant the chromosomes could not pair during meiosis.

- Chromosome mutation of the sterile hybrid led to a doubling in the number of chromosomes, making the hybrid fertile. This fertile hybrid is known as spelt wheat (*Triticum aestivum spelta*).

- Further gene mutation and selection has led to modern bread wheat (*Triticum aestivum*).

- More recently, plant breeders have improved the species by adding genes from other varieties and species. These have included genes to provide resistance to disease such as stem rust and fungal attack. Introduction of such genes involves crossing the high-yield variety with

> **Examiner's tip**
> You are unlikely to be asked to remember all the names of the different species involved in the evolution of modern bread wheat. They are supplied so that you are familiar with the names should they be given as part of a question.

another variety that shows the desired characteristic. The offspring are then tested to check that they show the desired combination of characteristics. If the desired combination is shown, the offspring are back-crossed with the high-yield variety for several generations to ensure that the offspring are pure-breeding.

Revision activity

Expand your mind map of variation by including ideas about natural and artificial selection and how evolution occurs.

Exam practice

1 **(a)** Explain what is meant by the term *locus*. [2]

    (b) **(i)** State the name given to the point at which chromatids break and rejoin. [1]

       (ii) Identify the stage of meiosis during which chromatids break and rejoin. [1]

      (iii) When the chromatids break and rejoin, genetic material can be exchanged. Describe the benefit to the species of this process. [4]

2 The following diagram shows a pair of homologous chromosomes at the start of meiosis. Two genes and their alleles are shown.

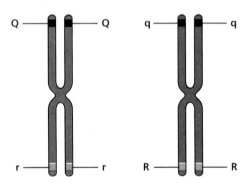

    (a) What is the genotype of this individual? [2]

    (b) Write down the possible gametes that could be produced, assuming that:

       (i) there are no crossovers [2]

      (ii) there is one crossover between the loci for genes r and q [4]

     (iii) there are two crossovers between the loci for genes r and q [2]

3 The following diagram shows the life cycle of a mammal.

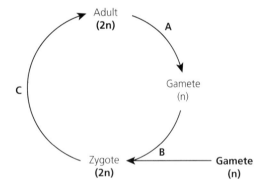

    (a) Identify at which stage (**A**, **B** or **C**) the following processes take place:

       (i) meiosis [1]

      (ii) mitosis [1]

     (iii) fertilisation [1]

    (b) Explain why the gametes contain the haploid (n) number of chromosomes. [3]

4   **(a)** Explain what is meant by the term *directional selection*.   [2]

    (b) Explain why variation is essential for selection to occur.   [2]

    (c) Using one example, describe how artificial selection can be achieved.   [5]

Answers and quick quiz 8 online

Online

Examiner's summary

By the end of this chapter you should be able to:

✔ Describe meiosis.

✔ Explain the terms *allele, locus, phenotype, genotype, dominant, codominant, recessive, linkage* and *crossing over*.

✔ Explain how meiosis and fertilisation lead to variation.

✔ Use genetic diagrams to solve problems involving sex linkage and codominance.

✔ Describe the interactions between loci (epistasis) and predict phenotypic ratios in problems involving epistasis.

✔ Use the chi-squared test.

✔ Distinguish between continuous and discontinuous variation.

✔ Explain how both genotype and environment contribute to phenotypic variation.

✔ Explain why variation is essential for selection.

✔ Use the Hardy-Weinberg principle.

✔ Explain how environmental factors can act as stabilising or evolutionary forces.

✔ Explain genetic drift.

✔ Explain the role of isolating mechanisms.

✔ Explain the various concepts of species.

✔ Compare and contrast natural selection and artificial selection and describe the use of artificial selection in producing modern dairy cows and bread wheat.

9 Cloning in plants and animals

Types of cloning

Reproductive and non-reproductive cloning

A clone is an exact, genetically identical copy of the parent. Cloning means to produce many genetically identical copies of individual organisms, cells or a single gene.

Reproductive cloning is the term used for the process of artificially cloning whole organisms. This is used in biotechnology and agriculture. **Non-reproductive cloning** is the process of making many identical copies without the intention of producing a new organism. This includes:

- making copies of a length of DNA by polymerase chain reaction (PCR) for use in research
- growing stem cells for research or the treatment of disease
- growing tissue in tissue culture for research or the replacement of tissues or organs

To be genetically identical, the cells and individuals must have been produced by cell division involving only mitosis.

> **Examiner's tip**
>
> This examination paper includes synoptic marks. These test your:
> - understanding of the principles behind different processes
> - ability to make links back to other parts of the specification
>
> The obvious links here are:
> - mitosis from F211
> - the production of food and evolution from F212

Cloning plants

Natural clones

Many plants reproduce asexually in nature. They do this by vegetative propagation, which includes:

- Producing runners or stolons, which are stems that grow along the surface or just below the surface of the ground. Occasionally these horizontal stems take root and grow a new vertical stem. Examples include strawberry and spider plants.
- Root suckers or basal sprouts, which are similar to runners but rather than growing a horizontal stem the plant produces new stems at intervals along the roots. These may form as a result of damage to the parent plant. Such suckers tend to grow in a circle around the old trunk, called a clonal patch. Elm trees produce root suckers and this has helped the species to survive Dutch elm disease, where the parent plant is killed by infection.
- Bulbs and tubers, which overwinter underground. Examples include patches of daffodils produced from one original bulb and the growth of many potato tubers from one original tuber.

Exam practice answers and quick quizzes at **www.therevisionbutton.co.uk/myrevisionnotes**

Artificial clones

Many plants can reproduce by fragmentation, which is when a small part of the plant regenerates a whole plant. Many gardeners will increase the number of plants in their garden by taking cuttings. This means cutting off a branch or stem and placing the cut end in the ground. Twigs cut from many bushes will grow roots and the success of these cuttings can be improved by dipping the cut ends in commercial rooting powder, which contains the plant growth regulator auxin.

Tissue culture is a modern development of this type of cloning. Much smaller pieces of plant tissue are placed on a special growth medium and caused to grow by mitotic cell division. They can be made to differentiate, forming new roots and leaves. This is also known as micropropagation and involves a number of steps:

1 The plant material is cut into small pieces called **explants**, which could be tiny pieces of leaf, stem, root or bud. The meristem (region of mitotically dividing cells) at the tip of the plant is often used because these cells are free from disease.

2 The explants are sterilised with bleach or alcohol to kill any bacteria and fungi. The following stages are then carried out under aseptic conditions.

3 The sterilised explants are placed on a growth medium (agar jelly) containing nutrients such as glucose, amino acids and phosphates. They also contain suitable high concentrations of plant growth substances such as auxin and cytokinin. This causes the cells in each explant to divide by mitosis to form a callus (a mass of undifferentiated cells that are all totipotent and therefore capable of giving rise to any cell).

4 Each callus can then be subdivided in order to increase the final number of plants made.

5 The callus is moved twice on to new growth media, each containing different ratios of plant growth substances. This makes the cells of the callus differentiate into tissues and organs to form a plant. For example, 100 auxin to 1 cytokinin causes roots to form and 4 auxin to 1 cytokinin causes shoots to form.

6 The tiny plantlets are then transferred to soil in a greenhouse, where they can acclimatise to normal growing conditions.

> **Tissue culture** is cloning cells from small groups of genetically identical cells to form a mass of similar cells.
>
> **Explants** are small pieces of tissue that can be grown to produce new plants.

> **Typical mistake**
>
> Candidates often forget that the callus can be further divided and that the procedure must be carried out aseptically.

> **Examiner's tip**
>
> The description of a technique such as this could be tested as part of How Science Works and you would be asked to justify certain steps. Alternatively, you could be asked to describe the process and a mark might be given for correctly sequencing the procedure.

> **Now test yourself**
>
> 1 Explain why it is essential to carry out tissue culture under aseptic conditions.
>
> Answer on p. 125
>
> Tested

Advantages and disadvantages

The advantages and disadvantages of plant cloning in agriculture are given in Table 9.1.

Table 9.1 The advantages and disadvantages of cloning in agriculture

Advantages	Disadvantages
It is a rapid way of producing new plants compared to growing plants from seed	Tissue culture is an expensive process because it is labour intensive
If tissue culture is done using the apical bud as an explant, the new plants will be disease-free (no viruses)	The tissue culture process can fail due to microbial contamination

Advantages	Disadvantages
Plants with specific characteristics can be selected (such as those producing high yields, showing resistance to a common pest or disease, or displaying a particular flower colour) and reproduced without forming hybrids that may have undesirable characteristics	The propagated clone may have a genetic disease or lack resistance to certain diseases
The new plants will be uniform — they all have the same phenotype. This makes growing and harvesting easier	All the cloned offspring are susceptible to the same pests or diseases. Therefore, if one is infected, they may all become infected. Crops are grown in monocultures, which allows pests and disease to spread easily
Infertile plants such as triploids can be reproduced, e.g. commercially grown bananas	There is little or no genetic variation. The only source of variation in a clone is mutation. Therefore, the gene pool is reduced and evolution is unlikely to occur
Plants that are hard to grow from seed can be reproduced, e.g. orchids	

Now test yourself

Tested ☐

2 Explain why it is important to avoid sexual reproduction if you want to create exact genetic copies.

Answer on p. 125

Revision activity

Draw a mind map with 'cloning plants' at the centre. Include natural and artificial cloning and the techniques used to clone plants, as well as their advantages and disadvantages.

Cloning animals

Natural clones

Revised ☐

Few animal species naturally form clones, but some invertebrate animals can reproduce asexually by budding or parthenogenesis. One notable exception is the production of identical twins in mammals (including humans). If a fertilised egg divides by mitosis and then the cells separate completely, two new individuals are produced. These two individuals are genetically identical.

Artificial clones

Revised ☐

It is possible to artificially produce clones in two ways: embryo splitting and adult cell cloning.

In embryo splitting, a zygote is created (usually by in vitro fertilisation or IVF). It is allowed to divide by mitosis to form a ball of cells, which are then separated before differentiation starts. Each cell is allowed to grow and divide again. Each new group of cells can then be placed into the uterus of a surrogate mother. This produces many genetically identical offspring. This technique can be used for reproductive cloning:

- to create many examples of elite farm animals that may have been produced by selective breeding or genetic modification (genetic engineering)
- to produce animals for research such as mice for testing new drug therapies

It can also be used for non-reproductive cloning to produce:

- genetically identical embryos from human or other animal cells for research into the action of genes that control development and differentiation
- new tissues and organs as replacement 'spare parts' for people who are ill. Tissues produced from a patient's own cells will be genetically identical and so will not be rejected by the immune system.

Adult cell cloning was first successful with Dolly the sheep in 1996. This technique is also called somatic cell nuclear transfer:

- An egg cell is harvested and its nucleus is sucked out using a micropipette.
- A normal body cell (somatic cell) from the adult to be cloned is harvested (in the case of Dolly, this cell was from the mammary gland of the original sheep).
- An electric shock triggers the enucleated egg cell and the somatic cell to fuse, starting mitosis and development as though it had just been fertilised.
- The embryo is placed into the uterus of the surrogate mother.

It is important to understand that this technique is not used commercially, just for research.

Advantages and disadvantages

Revised

The main problems with animal cloning are similar to the problems of plant cloning — a lack of genetic variability means that all the animals are equally susceptible to certain diseases, the gene pool is reduced and selection cannot take place. Also, cloned animals may be less healthy and have shorter life spans.

Table 9.2 The advantages and disadvantages of cloning animals

Advantages	Disadvantages
High-value animals — those giving high yields or good-quality meat — can be reproduced exactly	We are still not certain if the procedures involved cause other health issues and the animals produced may not be healthy in some way — they may be over-muscled or have some other deformity
Rare or endangered animals can be reproduced for conservation purposes	Some people are concerned about the welfare of the animals involved
Genetically modified animals can be reproduced quickly — if the animals are modified to create pharmaceutical chemicals, this helps produce the drugs quickly	Lack of genetic diversity means there is an inability to adapt to changes in the environment
	The exact genotype and phenotype of the organisms produced by embryo splitting depend on the sperm and egg used to create the zygote — the exact phenotype is not known until the cloned offspring are born
	The success rate of adult cell cloning is very poor and the method is a lot more expensive than conventional breeding

Revision activity

Draw a mind map with 'cloning animals' at the centre. Include natural and artificial cloning and the techniques used to clone animals as well as their advantages and disadvantages.

Exam practice

1 **(a)** **(i)** Explain what is meant by the term *non-reproductive cloning*. [2]

   (ii) Describe two uses for non-reproductive clones. [2]

   (b) **(i)** Explain what is meant by the term *reproductive cloning*. [1]

   (ii) Describe three advantages of plant tissue culture. [3]

   (iii) Dolly the sheep was cloned by adult cell cloning. Outline the technique used to clone a mammal from an adult cell. [4]

2 **(a)** The usual method for growing potatoes is to save some of the harvested crop and replant them the following spring.

   (i) State two advantages of growing potatoes in this way. [2]

   (ii) In Ireland, potato blight caused by the fungus *Phytophthora infestans* caused widespread destruction of potato crops between 1845 and 1850. Suggest what made the potato crop so susceptible to the fungus. [3]

   (b) Modern potato stocks can be certified free from disease.

   (i) Describe how a blight-resistant potato plant could be produced. [3]

   (ii) Describe how large numbers of resistant plants could be produced quickly. [4]

Answers and quick quiz 9 online

Online

Examiner's summary

By the end of this chapter you should be able to:

✔ Outline the differences between reproductive and non-reproductive cloning.

✔ Describe the production of natural clones in plants using the example of vegetative propagation in elm trees.

✔ Describe the production of artificial clones of plants from tissue culture.

✔ Discuss the advantages and disadvantages of plant cloning in agriculture.

✔ Describe how artificial clones of animals can be produced.

✔ Discuss the advantages and disadvantages of cloning animals.

10 Biotechnology

Using microorganisms

Examiner's tip

This examination paper includes synoptic marks. These test your:
- understanding of the principles behind different processes
- ability to make links back to other parts of the specification

The obvious links here are:
- enzyme action, the conditions needed for enzymes to work well and protein synthesis from F212
- respiration from F214

Defining biotechnology   Revised

Biotechnology is the industrial use of microorganisms to carry out processes such as:

- producing food — whole microorganisms are used in making cheese, yoghurt, bread, beer and single-cell proteins; enzymes are used in other processes, such as pectinase to release juice from fruit and glucose isomerase to make sweet products

- producing medicinal drugs — whole organisms are used in manufacturing drugs such as penicillin and insulin; genetically modified mammals such as sheep and goats are also used to produce useful proteins, which are harvested in their milk

- producing other products such as methane or biogas; citric acid, which is the food preservative E330

- treating waste products, such as waste water treatment and sewage treatment

Biotechnology is the use of organisms or parts of organisms (e.g. enzymes) in industrial processes.

Examiner's tip

Biotechnology is still a growth area and more applications are continually found. This gives plenty of scope to test your understanding through using unfamiliar examples, so be ready to think about the example given and apply the principles.

Why are microorganisms used?   Revised

Microorganisms are not the only organisms used in biotechnology. However, microorganisms such as bacteria and fungi are often used because they:

- grow rapidly, even at low temperature
- reproduce quickly
- reproduce asexually so that all individuals are genetically identical
- remain as single cells or small clumps, which means that the cells do not differentiate and all the cells remain productive
- can produce proteins or other substances that are secreted or excreted to their external environment and can be isolated relatively easily in a pure form, which means lower downstream processing costs

- can carry out quite complex processes that would be difficult using chemicals
- can be genetically engineered
- can be selected for certain characteristics easily (strain selection)
- can be grown or used anywhere, making the process independent of climate
- can often be grown on waste products from other processes, e.g. molasses

Revision activity

Write a list of all the topics here that link to other parts of the specification you have learnt about.

Standard growth curves

Revised

The **standard growth curve** for a microorganism population is called a sigmoid curve. In a **closed culture** it has four phases (Figure 10.1):

1 Lag phase — reproduction and growth are very slow as cells acclimatise, absorb nutrients, produce enzymes and store energy. This may involve activating genes to produce specific enzymes.

2 Exponential (log) phase — there is a rapid reproduction rate and cells may divide every 20–30 minutes so the population could double with every generation. There are no limiting conditions and few cells die.

3 Stationary (stable) phase — the population remains constant as the death rate equals the reproduction rate.

4 Decline phase — the death rate exceeds the reproduction rate and the population may decline to zero, often because there is some limiting condition such as:
  - high temperature
  - limited nutrients or oxygen
  - a build-up of waste products such as carbon dioxide or ethanol

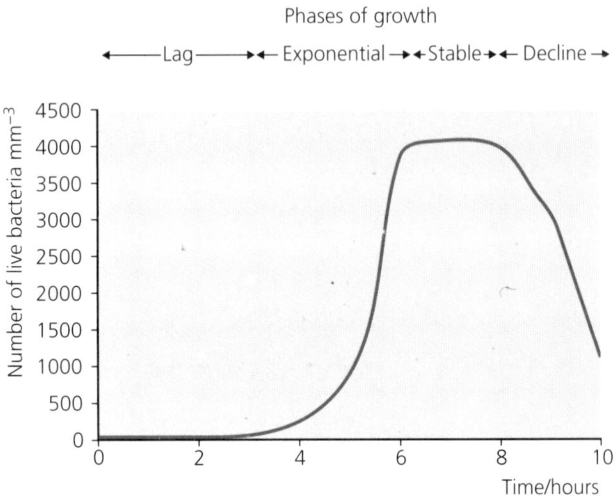

Figure 10.1 The changes in numbers of bacteria in pure culture. The lag phase to the stationary (stable) phase is called sigmoid growth

Typical mistake

Some candidates confuse a standard growth curve showing population size with a rate of growth curve.

Revision activity

Sketch a growth curve and annotate the graph with notes that explain the shape.

Now test yourself

Tested

1 Explain why the standard growth curve comes to a plateau.

Answer on p. 125

Using enzymes

Immobilised enzymes

Revised

Immobilised enzymes are enzymes that are fixed in place so that they do not mix freely with the substrate and are not lost in the process. When a single-step process is required, it may not need the whole organism to be present. Enzymes alone can be used in many such industrial processes and there are several ways to immobilise enzymes:

- adsorption on to a surface such as glass, carbon or clay — the enzymes are held in place on the surface by hydrophobic or ionic interactions, or possibly by covalent bonding
- entrapment in gel beads or between fibres of cellulose
- separation from the substrate by a partially permeable membrane

Immobilised enzymes have many advantages in **large-scale production**:

- Enzymes are good catalysts.
- Enzymes are specific to the process, so there are few by-products that would need to be removed from the final product.
- Reactions can be carried out at fairly low temperatures around 30–70°C, which saves heating costs.
- The enzyme is kept separate from the product so that they do not contaminate the end product.
- The enzymes are not lost and can be reused straight away.
- The enzymes are often protected by the immobilising matrix so that they are less likely to be damaged by high temperature or extremes of pH.

> **Revision activity**
>
> Draw a mind map with 'immobilised enzymes' at the centre. Include all the information about how enzymes can be immobilised, enzyme activity and the effects of surrounding conditions.

Growing conditions

Continuous and batch culture — Revised

Microorganisms can be grown in two ways:

- **Continuous culture** is where a culture is set up, and nutrients are added and products removed from the culture at intervals. The culture is maintained at the exponential phase of the standard growth curve so that it continues to grow and produce its **metabolites** quickly.
- **Batch culture** is where a starter population of microorganisms is supplied with a fixed amount of nutrients and allowed to grow. At the end of the time period, the products are extracted.

> **Metabolites** are substances produced by living organisms.

Table 10.1 Comparing continuous and batch culture

Continuous culture	Batch culture
Nutrients are added continuously	Nutrients are added at start only
Maintaining the correct conditions can be difficult and expensive	The culture can be left to continue for a set time period
There is a high growth rate as nutrient levels are maintained	There is a slower growth rate as nutrients decline
The products are collected continuously	The products are collected at the end
The fermenter is in use continuously, so this is more efficient	The process is less efficient as the fermenter is not in use constantly
The microorganisms are metabolising normally, which is useful for producing primary metabolites	This is useful for secondary metabolites, as these are generated when population growth rate declines
Examples include production of insulin and single-cell protein	Examples include wine, beer and yoghurt
In the event of contamination, production is stopped and losses can be great	In the event of contamination, only one batch is lost

Primary and secondary metabolites
Revised

Primary metabolites are substances produced by living organisms in the course of normal metabolism. All microorganisms produce primary metabolites. Examples include proteins, enzymes and alcohol.

Secondary metabolites are substances produced by living organisms after the main population growth has occurred and they are not essential to the survival of the organism. Not all microorganisms produce secondary metabolites. They are usually produced when nutrients are in short supply and the population is not growing rapidly. Examples include antibiotics.

Typical mistake

Try not to confuse similar terms such as *primary metabolites* and *secondary metabolites, continuous culture* and *batch culture*.

Now test yourself

2   Explain why batch fermentation is best for harvesting secondary metabolites.

Answer on p. 125

Tested

Manipulating growing conditions
Revised

Microorganisms are grown in **fermentation vessels** called fermenters, which can be used to optimise **growing conditions** in order to maximise the yield of the required product. The following conditions are maintained or even altered during the fermentation process:

- Temperature — it needs to be warm enough to maintain the metabolic rate but not hot enough to denature the enzymes. Once the fermentation is in process, the microorganisms produce their own heat and it may be necessary to cool the fermentation vessel with cold water.

- Substrate or nutrients — the correct nutrients must be added, including sources of carbon and nitrogen. When these are added depends on the type of culture (batch or continuous) and whether a primary or secondary metabolite is required.

- Oxygen — this is required when processes involving aerobic respiration are being used, but not if the product is achieved through anaerobic respiration (e.g. alcohol).

- pH — this must be maintained to ensure enzyme action continues as required.

- Agitation — if the culture is not stirred or mixed in some way, the microorganisms may settle to the bottom of the fermenter, which would reduce activity. The culture can be agitated and kept moving by a rotating paddle or by bubbling oxygen in at the base.

Examiner's tip

Remember that enzymes are involved, so all conditions need to be suitable for enzymes and there is an obvious link for testing your knowledge synoptically.

Revision activity

Sketch a diagram of a fermenter and annotate it with notes explaining the need for each feature.

Now test yourself

3   Suggest how continuous culture could be used to produce secondary metabolites such as antibiotics.

Answer on p. 125

Tested

Asepsis
Revised

Nutrients, oxygen and an ideal temperature are provided in a fermenter so that the microorganisms are growing in ideal conditions. It is important to exclude other microorganisms as they would also grow well. Unwanted microorganisms could:

- compete for nutrients, reducing growth rate and yield
- kill the culture microorganism
- use the required product for their own metabolism
- spoil the product by contaminating it with toxic by-products

A contaminated batch would have to be thrown away, losing the industrial company money. **Asepsis** involves avoiding contamination by other microorganisms. It is achieved by:

> **Asepsis** is keeping the culture free from unwanted microorganisms.

- washing, disinfecting and steam cleaning all equipment, including the fermenter — this takes time between batch cultures, but only needs to be done once before a continuous culture is started
- using a fermenter made from polished stainless steel, which prevents microbes sticking to surfaces
- sterilising all nutrients before they are added to the fermenter, which is usually achieved by steam or heat treatment
- making sure that any air or oxygen bubbled into the fermenter is free from microorganisms, which is achieved using very fine filters

Now test yourself

4 Explain why a fermenter must be cleaned and disinfected thoroughly between batches.

Answer on p. 125

Tested

Exam practice

1 High-fructose corn syrups are produced using an immobilised enzyme called glucose isomerase.
  (a) Suggest how an enzyme can be immobilised for student investigation. [2]
  (b) Describe two advantages of immobilising enzymes in an industrial process. [2]
  (c) Fructose and glucose are both monosaccharides with the formula $C_6H_{12}O_6$. Suggest what is the action of glucose isomerase. [2]
2 (a) (i) Explain what is meant by the term *fermenter*. [2]
      (ii) State two conditions inside a fermenter that are controlled and explain why it is necessary to control each. [4]
      (iii) Describe how a fermenter is kept aseptic. [4]
  (b) Compare and contrast the production of primary and secondary metabolites in a fermenter. [7]

Answers and quick quiz 10 online

Online

Examiner's summary

By the end of this chapter you should be able to:
- ✔ State that biotechnology is the industrial use of living organisms (or parts of living organisms) to produce food, drugs or other products.
- ✔ Explain why microorganisms are often used in biotechnological processes.
- ✔ Describe, with the aid of diagrams, and explain the standard growth curve of a microorganism in a closed culture.
- ✔ Describe how enzymes can be immobilised.

- ✔ Explain why immobilised enzymes are used in large-scale production.
- ✔ Compare and contrast the processes of continuous and batch culture.
- ✔ Describe the differences between primary and secondary metabolites.
- ✔ Explain the importance of manipulating the growing conditions in a fermentation vessel in order to maximise the yield of product required.
- ✔ Explain the importance of asepsis in the manipulation of microorganisms.

11 Genomes and gene technologies

Genome sequencing

Examiner's tip

This examination paper includes synoptic marks. These test your:
- understanding of the principles behind different processes
- ability to make links back to other parts of the specification

The obvious links here are:
- the idea of complementary shapes of molecules from F211 and F212
- enzyme action and the conditions needed for enzymes to work well, DNA structure and improving the production of food by natural and artificial selection from F212

Sequencing genomes
Revised

A **genome** is all the genetic information inside a cell or organism. The genome includes coding DNA (consisting of genes that code for the structure of polypeptides) and non-coding DNA (which carries out a range of regulatory functions). **Sequencing** a genome involves a series of steps:

1 The DNA is extracted from the cells.

2 It is cut into sections of about 100 000 nucleotides in length using **restriction endonuclease** enzymes.

3 These sections are placed in bacterial artificial chromosomes (BACs) and transferred to cells of E. coli.

4 These E. coli cells are grown in culture to clone many copies of the DNA.

5 The DNA is extracted from the E. coli.

6 The DNA is cut into shorter fragments (of up to 1000 nucleotides) using a range of different **restriction enzymes**. This ensures that the sections are cut in a variety of different places so that the fragments overlap rather than producing many fragments of the same length.

7 The fragments are separated by gel **electrophoresis**.

8 Each fragment is sequenced using an automated process.

9 The overlapping fragments are compared to reassemble the whole BAC sequence.

The automated sequencing process

Lengths of up to 1000 nucleotides can be sequenced by computers. An **isolated DNA fragment** is copied using the **polymerase chain reaction (PCR)**. However, along with normal nucleotides, the reaction

> A **restriction endonuclease** is an enzyme that cuts DNA into sections.

Examiner's tip

You are only expected to know an outline of this process — there is no need to confuse it with too much detail.

Revision activity

Draw a flow chart to represent the sequencing process.

Now test yourself

1 Explain why it is essential to use a range of different restriction endonucleases.

Answer on p. 125

Tested

vessel is supplied with stained nucleotides, which stop the growth of the new DNA strand. Therefore, the reaction vessel contains a wide range of lengths of newly made DNA fragments, each ending in a modified nucleotide with a stain. The mixture is separated by electrophoresis in a capillary tube. The shortest lengths move more quickly and are detected as they move past a detector first, followed by fragments of increasing length. The detector reads the stain on each fragment and produces a sequence.

Using genome sequences
Revised

The sequence of nucleotides in the genome or in a particular gene allows for **genome-wide comparisons** with the sequence in another individual or species. This can be useful to:

- ascertain how important the gene is — it may code for an essential protein or simply for an additional feature in one species
- work out evolutionary relationships — the more similar the sequence, the more closely related the species
- compare the genome of a pathogenic organism with that of a similar non-pathogenic organism — this enables us to find out what genes or base sequences are important in causing the disease, which could help in the design of drugs to combat the disease
- identify which genes cause inherited diseases by comparing the base sequence of a healthy person with that of someone who has an inherited disease
- investigate the effects of specific mutations by manufacturing a mutation and inserting it into a microorganism such as yeast

Revision activity

Draw a mind map with 'genome sequence' at the centre. Don't forget to include DNA structure, the genetic code and protein synthesis as synoptic topics.

Now test yourself

2 Explain how a genome sequence can be used to work out evolutionary relationships.

Answer on p. 125

Tested

Genetic manipulation

Recombinant DNA and genetic engineering
Revised

Recombinant DNA is DNA from one species that has been modified by incorporating one or more genes from another organism, often from another species. It is produced by **genetic engineering**, which is also known as recombinant gene technology. There are a number of techniques involved:

1 The desired gene is extracted.
2 The gene is isolated from other DNA using electrophoresis and probes.
3 Multiple copies of the gene are made by the polymerase chain reaction (PCR).
4 The gene is placed in a **vector**.
5 The vector is used to insert the gene into the recipient.
6 The **receiving organism** uses the gene to produce the desired product or characteristic.

Recombinant DNA is DNA from one organism that contains a gene from another organism or species.

A **vector** is a means of inserting DNA into a cell.

Typical mistake

Remember the difference between the gene and the gene product — many candidates confuse the two.

Extraction of genes   Revised

Restriction endonuclease enzymes are specific to DNA. There is a wide range of endonucleases, each with a different active site that is specific to a particular sequence of nucleotides (the restriction site). Scientists can now make restriction enzymes that cut at specific nucleotide sequences — each specific enzyme always cuts DNA at the same sequence of nucleotides. The enzyme hydrolyses the bonds along the sugar–phosphate backbone. The DNA is cut in a staggered fashion, leaving a sequence of a few nucleotides unpaired and exposed. This is called a sticky end as it can be used to join to another length of DNA with the complementary sequence of exposed nucleotides. An alternative way to get the gene is to extract mRNA from the pancreatic cells and use reverse transcriptase to manufacture the gene.

Typical mistake

Take care with your wording here. Many candidates say 'endonucleases cut the gene', but this is not correct. They cut the DNA into sections.

Electrophoresis   Revised

Electrophoresis is the technique of separating pieces of DNA according to their length. It is similar to chromatography, but a gel substrate is used in place of paper or silica. The gel contains wells into which the sample of DNA is placed and is covered by a buffer solution to prevent it drying out. The DNA fragments are free to diffuse through the gel, but they diffuse slowly because of their size. They can be made to move more quickly and in one direction by application of a potential gradient along the gel. DNA is negatively charged and the fragments move towards the positive electrode end of the gel. The shorter lengths of DNA are smaller and travel more quickly. Therefore, the fragments of different lengths are separated.

Electrophoresis is the technique used to separate lengths of DNA by their size.

DNA probes   Revised

The DNA sequence is not visible, so it needs to be labelled in some way so that scientists can identify the desired gene or sequence of nucleotides. The DNA can be identified using a **DNA probe** or, more specifically, a **gene probe**. Probes are useful in locating a specific gene desired for genetic engineering, to compare the genome of different species or to carry out genetic screening (to see if an individual has a specific genetic disease).

A gene probe is a single-stranded short length of DNA. It has a specific sequence that complements a sequence in the desired fragment of DNA (possibly a whole gene). It is also labelled in some way:

A **gene probe** is a short length of DNA that is used to identify genes with a specific nucleotide sequence.

● The nucleotides can be made using 32P, which shows up on a photographic plate.
● It could be attached to a specific stain.
● It may be a fluorescent molecule that glows in UV light.

Probes can be used in different ways:

● A suspension of probes can be poured on to the gel electrophoresis plate after the DNA fragments have been separated. This allows the probes to adhere to the complementary sequence and identifies where on the electrophoresis plate the desired gene can be found.

- The probes can be mixed with the fragmented DNA in suspension, where they combine with the desired complementary sequence. The fragments of DNA can then be separated by gel electrophoresis or centrifugation.
- Probes can also be used in a similar fashion to immobilised clonal antibodies. A specific probe is immobilised on a surface and fragments of DNA are allowed to pass over the probes. If the desired DNA sequence is present, the DNA sticks to the probe.

The polymerase chain reaction   Revised

The polymerase chain reaction is a chemical reaction that replicates DNA. Sequencing or identifying specific DNA requires a large number of molecules. If only a small sample is available, it can be amplified (replicated many times to give multiple copies). This is achieved by the polymerase chain reaction (PCR). The sequence of steps in PCR is as follows:

1 Mix the sample of DNA with extra DNA nucleotides and DNA polymerase.
2 Heat to 95°C, which breaks the hydrogen bonds holding the two strands together and splits the double helix to form single strands with exposed bases.
3 Cool to 55°C and add short lengths of single-stranded DNA known as primers, which bind to the strands to be replicated. This produces double-stranded sections that allow the DNA polymerase to bind.
4 Raise temperature to 72°C, which is the optimum temperature for this DNA polymerase (Taq polymerase).
5 The DNA polymerase adds nucleotides to the double-stranded section until complete new double-stranded molecules are created.
6 Repeat the sequence many times.

Now test yourself

3 What is special about Taq polymerase?

Answer on p. 125

Tested

Placing isolated DNA into plasmids   Revised

Once a gene or length of DNA has been isolated, it must be inserted into the recipient cell, which is done using a vector. Bacteria contain short circular sequences of DNA called **plasmids**, which can be used as vectors. They can be modified or created and placed into bacteria as follows:

Plasmids are short circular pieces of DNA.

1 The gene is isolated using a restriction endonuclease, which cuts the gene out of the DNA so that it has sticky ends.
2 If a plasmid is cut using the same restriction endonuclease, it has the same sticky ends as the gene.
3 Mixing the isolated gene with the cut plasmids allows some plasmids to take up the gene and join up to reform a full circle. This is called annealing.
4 **Ligase** enzymes reform the sugar–phosphate backbone of the DNA.
5 Some of the bacteria growing in a culture medium containing the newly formed plasmids take up the plasmids (not all do this — only about 0.25%).
6 Heat-shock and the presence of calcium salts can encourage the bacteria to take up the plasmids — the culture is cooled to 0°C and then quickly heated to 40°C.

7 Those bacteria that have taken up the plasmids are called **transgenic microorganisms**.

8 These transgenic microorganisms express the **desired gene product**.

Bacteria and other microorganisms often reproduce asexually. This means that there is no genetic variation introduced by meiosis and random fertilisation. Taking up DNA from their surroundings introduces genetic variation and increases diversity. This may enable the cell to survive in its environment or to take advantage of certain conditions and thrive where others struggle to survive. Selection and evolution become possible.

Examiner's tip

Here the endonucleases are used to cut the plasmid open, not to cut sections out of it.

Now test yourself

4 Explain why sticky ends are essential.

Answer on p. 125

Tested

Other vectors

Revised

Viruses can also be used as vectors. The DNA placed into a virus will be inserted into the recipient cell when the virus attacks the cell. Ti plasmids can be inserted into bacteria (*Agrobacterium tumefaciens*), which then infect plants. The bacterium inserts the plasmid into the plant DNA. Bacterial artificial chromosomes (BACs) are circular lengths of DNA used to clone genes inside bacteria, whereas yeast artificial chromosomes (YACs) do the same in yeast.

Liposomes are tiny balls of fatty molecules that can fuse to cell surface membranes or even cross lipid bilayers. The gene is held inside the ball of fatty molecules and enters the cell as fusion occurs.

Genetic markers

Revised

A **genetic marker** is a sequence of DNA used to test that the required gene has been inserted successfully. Only about 0.25% of cultured bacteria take up the plasmid containing the desired gene. It is important to isolate those that have the plasmid from those that do not, as it is pointless to culture the bacteria that will not be able to express the gene and produce the required product. This is achieved using genetic markers, as in this example:

1 The gene is not introduced on its own.

2 The length of DNA inserted into plasmids also contains genes for resistance to two antibiotics (tetracycline and ampicillin).

3 The desired gene is inserted into the middle of the gene for resistance to tetracycline.

4 Therefore, any bacteria that have taken up the plasmid successfully will be resistant to ampicillin but not to tetracycline.

5 The bacteria are tested to see if they are resistant to the two antibiotics by replica plating:
   ● Grow the bacteria on normal nutrient agar.
   ● Blot with sterile velvet to pick up the cells from each colony and transfer to agar containing ampicillin. Only the bacteria with the plasmid containing the ampicillin-resistance gene will grow.
   ● Blot again and transfer the cells to agar containing tetracycline. Only the cells with a complete tetracycline-resistance gene will grow.
   ● The cells that grew on the ampicillin but not on the tetracycline are the ones that contain the required gene.

Revision activity

Draw a mind map with 'genetic engineering' at the centre. Include each of the techniques covered above to show the links between the techniques.

Now test yourself

5 Explain why the bacteria that have successfully taken up the plasmid will be resistant to one antibiotic but not to the other.

Answer on p. 125

Tested

Examples of genetic engineering

Revised

Human insulin

Human insulin can be produced by engineering bacteria to contain the gene and express it:

1 Extract mRNA from cells of a human pancreas. Use reverse transcriptase to manufacture the gene for insulin.

2 Insert the gene into a bacterial plasmid using ligase enzyme, along with markers.

3 Allow bacteria to take up the plasmid.

4 Grow on nutrient agar.

5 Use the marker genes to check which bacteria have taken up the plasmid.

6 Isolate those bacteria that have the human insulin gene.

7 Grow in a fermenter using continuous culture.

Examiner's tip

Look for synoptic links back to F214 and the treatment of diabetes.

Golden Rice

Golden Rice is rice that contains beta-carotene, which is used to make vitamin A in the human gut. It was produced by inserting two genes into rice:

1 Isolate the genes for phytoene synthetase and Crt1 enzyme. The phytoene synthetase gene was extracted from daffodils and the Crt1 gene was extracted from bacteria (*Erwinia uredovora*).

2 Insert these genes into the rice genome close to the area that promotes growth of the endosperm (seed food store).

3 The transgenic rice plant develops new seeds (rice grains) that have beta-carotene in the endosperm.

4 Cross-breeding with other rice varieties increases the yield.

Xenotransplantation

Xenotransplantation is the transplant of organs from one species to another. Certain species have organs of similar size to human organs — notably pigs are very similar. Pigs and sheep can be engineered to make xenotransplantation easier. Genes for the cell surface proteins found on human cells can be inserted into the animal embryo. It is also possible to cause mutations that prevent the expression of the original antigens so that the animal produces cells with human proteins and not pig or sheep proteins on their cell surface membranes.

Gene therapy

Revised

Gene therapy is treating genetic disorders by inserting new genes. It does not involve replacing genes — it is adding to the genome of a cell. If a functioning gene is inserted into a cell without a functioning gene, the new gene can be expressed.

Somatic cell gene therapy is adding genes into a body cell so that the cell can produce a specific protein. This allows individual cells or tissues in an organism to be augmented. Cells can also be killed in this way by inserting a gene that enables it to manufacture a foreign antigen. This allows the cell to be recognised as foreign so that it will be attacked by the immune system, which can be useful to treat cancer. Treating somatic cells (body cells) is short-term and affects only the cells directly treated.

Germ line gene therapy is treating a fertilised egg, which means that all cells in that organism will possess the required gene that will be passed on to its offspring.

Ethical concerns

Revised

Genetic engineering is a powerful new technology with huge potential to improve food production and medicine and to reduce human suffering from disease and starvation. However, genetic manipulation of organisms causes many potential **ethical concerns**, including:

- the possible spread of antibiotic-resistance genes used as markers to 'wild' bacteria
- the reduction of genetic diversity if only certain genetically modified (GM) crops are grown
- the spread of genes for resistance to pests or herbicides from crops to weeds, making 'superweeds'
- causing suffering to farm animals if the genes introduced adversely affect their metabolism
- the possible presence of toxins or allergens in GM crops
- the concern that, as a new technology, we simply do not know what potential risks may exist

Now test yourself

6 Explain why there are more concerns about the use of germ line gene therapy than somatic cell gene therapy.

Answer on p. 125

Tested

Exam practice

1 **(a)** Name the following:

    (i) enzymes that synthesise new DNA   [1]

    (ii) enzymes that cut DNA at specific sequences   [1]

    (iii) enzymes that seal two pieces of DNA together   [1]

    (iv) small circular pieces of DNA in bacteria   [1]

(b) Suggest a suitable vector for each of the following genetic engineering procedures:

    (i) transforming bacteria to express the gene for human insulin   [1]

    (ii) engineering rice plants to express the enzyme to make vitamin A   [1]

    (iii) treating cystic fibrosis by introducing working alleles to the lung cells of a human patient   [1]

    (iv) cloning a large human gene inside bacteria   [1]

(c) The following list represents steps in the genetic engineering process to make bacteria capable of producing human insulin. Place the steps in the correct order.   [5]

    A Use reverse transcriptase to make DNA and add sticky ends.

    B Incubate a plasmid with a restriction enzyme.

    C Extract mRNA from human pancreatic cells.

    D Heat-shock the bacteria in the calcium chloride solution and add the recombinant vector.

    E Incubate the prepared gene with a cut plasmid.

2 (a) Haemoglobin is made up of the polypeptides α-globin and β-globin. Digestion of normal human DNA with the enzyme HPa I produces a fragment 7.6 kbp long containing the β-globin gene. Digestion of DNA from a person with sickle-cell anaemia produces a fragment 13 kbp long.

    (i) What type of enzyme is HPa I? [1]

    (ii) Describe how these fragments could be separated using gel electrophoresis. [4]

(b) (i) What is a gene probe? [2]

    (ii) Explain how a gene probe can be used to screen members of a family with a genetic disorder such as muscular dystrophy. [5]

3 (a) During the polymerase chain reaction, the temperature of the mixture is altered. List the temperatures involved and describe the process occurring at each temperature. [6]

(b) Explain the purpose of adding primers to the PCR mix. [3]

Answers and quick quiz 11 online

Online

Examiner's summary

By the end of this chapter you should be able to:

- Outline the steps involved in sequencing the genome of an organism and how gene sequencing allows for genome-wide comparisons between individuals and between species.
- Define the term *recombinant DNA*.
- Explain what is meant by genetic engineering including extraction of DNA, separation using electrophoresis, use of DNA probes, the polymerase chain reaction (PCR), use of vectors such as plasmids and how plasmids may be taken up by bacterial cells.
- Describe the advantage to microorganisms of the capacity to take up plasmid DNA from the environment.

- Outline how genetic markers in plasmids can be used to identify the bacteria that have taken up a recombinant plasmid.
- Outline the processes involved in the genetic engineering of bacteria to produce human insulin, the genetic engineering of Golden Rice and how animals can be genetically engineered for xenotransplantation.
- Explain the term *gene therapy* and the differences between somatic cell gene therapy and germ line gene therapy.
- Discuss the ethical concerns raised by the genetic manipulation of animals (including humans), plants and microorganisms.

12 Ecosystems

Ecosystems as complex interactions

What are ecosystems?   Revised ☐

An **ecosystem** is all the living things in one area, their interactions with each other and their interactions with their environment including factors in the soil. Environmental factors are:

- **biotic factors** — those that involve other living things, such as predation, food supply, disease and competition
- **abiotic factors** — those that involve non-living things, such as pH, temperature, moisture/rainfall and soil type

Ecosystems are not static; they are always changing. They are complex with so many interlaced interactions that one change, however small, will cause other changes. Therefore, ecosystems are **dynamic systems**. Any factor that affects the way an organism lives or makes food easier to access or more difficult to find will cause some change in the ecosystem. Such changes alter the flow of energy through the ecosystem.

> An **ecosystem** describes all the interactions between the living things and the non-living components in a habitat.

> **Typical mistake**
>
> Many candidates forget about the abiotic factors when they are describing an ecosystem.

Trophic levels   Revised ☐

An ecosystem is organised into **trophic levels**, which are feeding levels:

- **Producers** are the lowest trophic level — they are autotrophs. They convert energy from the environment (and inorganic matter) into chemical energy in the form of complex organic molecules. These molecules are then used for growth or as substrates in respiration. Producers include:
  - plants, known as phototrophs, which use the sun's energy to convert small inorganic molecules into carbohydrates
  - certain bacteria, known as chemotrophs, which can use chemical energy and heat to convert small inorganic molecules into complex organic molecules

- **Consumers** occupy the higher trophic levels. These are organisms that feed on other organisms. They digest the complex organic molecules made by autotrophs and then use the products for growth or as substrates in respiration:
  - Primary consumers feed on plants.
  - Secondary consumers feed on primary consumers.
  - Tertiary consumers feed on secondary consumers.
- **Decomposers** are organisms that feed on waste or dead organic matter. They gain their energy by digesting and respiring the complex molecules in the organic matter. They cause decay and food to go off.

Revision activity

Draw a food web of a simple ecosystem — remember to keep organisms in trophic levels. At the side of each trophic level, explain what role the organisms in the trophic level play in the ecosystem.

Energy transfer through ecosystems

Energy transfer
Revised

In most ecosystems, energy comes from the sun. It enters the ecosystem in the form of light and is trapped by the producers (green plants) in the process of photosynthesis. The plants use the energy to manufacture carbohydrates that are used for growth. The energy is then transferred from one trophic level to the next by feeding.

Measuring energy transfer
Revised

The energy in a trophic level can be measured by calculating the amount of energy held in the organic matter of the living organisms in that trophic level. This can be achieved by sampling the organisms and measuring how much energy is held in each organism or in each kilogram of body mass. This must be multiplied by the number of organisms or the number of kilograms in the trophic level.

A bomb calorimeter is used to measure the amount of energy held in the organism. This is a device that burns the body quickly in a good supply of oxygen and uses the energy released to heat water.

Typical mistake

Many candidates find even simple calculations difficult. They are particularly unsure what to put as the 'energy in trophic level 1' figure in this formula.

The efficiency of energy transfer is then calculated using the following formula:

$$\text{Efficiency of transfer} = \left(\frac{\text{energy in trophic level 2}}{\text{energy in trophic level 1}} \right) \times 100$$

Efficiency of energy transfer
Revised

The transfer of energy from the sun to plant growth is not very efficient. Only about 1% of the sun's energy reaching the ground is converted into plant growth. This is because:

- some light falls on bare ground
- some light passes through the leaf
- some light is reflected, which is why leaves look green
- some light may be the wrong wavelength to be useful

- some of the energy is heat and cannot be used in photosynthesis
- many plants do not have leaves all year round

The amount of energy converted to complex molecules is called the gross primary productivity. Some of these molecules are used in respiration and do not contribute to growth — the growth is known as the net primary productivity. Net primary productivity = gross primary productivity – respiration.

The transfer of energy between trophic levels is better. As a general rule, about 10% of the energy in one trophic level enters the next trophic level. This is because of energy losses at each level and between levels. Losses include:

- energy released in respiration to power life processes such as movement and active transport
- tissue that cannot be eaten or digested easily such as cellulose and bone
- energy held in the substances excreted
- organisms that grow but are not eaten by organisms in the next level

The losses between trophic levels vary between ecosystems and some are more efficient at passing energy between trophic levels than others. Reasons for this include the following:

- Ectotherms use less energy in keeping warm than endotherms, so more of the energy consumed can be used in growth and is available to the next trophic level.
- In some ecosystems a higher proportion of the available food is eaten.
- Some tissues are easier to digest than others.

Revision activity

Sketch a leaf and show energy coming from the sun. Use arrows to indicate what happens to the energy and explain what is represented by each arrow.

Examiner's tip

Remember that net productivity is gross productivity less losses due to respiration.

Human activities   Revised ☐

Humans have learnt to manipulate the flow of energy through ecosystems to improve the growth of food. Principally, this involves diverting as much energy into the growth of food as possible and creating artificial ecosystems. This involves:

- improving the conditions for growth — using greenhouses where crops can be grown in artificial conditions to enhance their growth (extra light, warmer temperatures and extra carbon dioxide will all help); irrigation, fertilisers and crop rotation also help to improve growth
- improving the efficiency of energy conversion to food — using faster-growing crops; selective breeding or genetic engineering to produce fungus resistance or disease resistance in plants or faster growth/greater production from animals; harvesting animals as soon as they are fully grown; use of steroids to increase muscle (meat) growth; keeping animals in small enclosures or even indoors to reduce loss from respiration used to move around and keep warm
- reducing competition — control of pests; use of selective weedkillers; use of antibiotics to treat illness

Ecosystems as dynamic entities

Succession

Succession is used to describe the way that a community changes over time. Each new species living in a habitat modifies that habitat, changing the conditions slightly. As a result, new species may be able to migrate in and live successfully. As the conditions change, species that have lived there successfully may be outcompeted by newer arrivals. Therefore, the community living in that habitat changes over time. As a general rule, the community becomes more complex as succession continues and the organisms that are able to survive tend to get larger.

Primary succession is the sequence of organisms that live in an area starting from a newly created piece of land that has not been occupied previously. Such succession starts with pioneer species that are hardy and able to survive in harsh conditions. These colonisers are usually small and few in number. The succession continues as each new arrival modifies the environment and the community becomes more complex with larger numbers of larger species. It continues until the **climax community** (the final community full of well-adapted competitors) is reached and can remain unchanged for many years. One example of primary succession is seen on sand dunes:

> bare sand — sea rocket — marram grass — bird's foot trefoil — clover — grasses — small shrubs (bramble) — small trees (elder) — larger trees (silver birch)

> **Succession** is a directional change in the composition of a community.

> **Examiner's tip**
>
> You do not need to remember a large number of examples of succession. Learn one and remember that succession starts with pioneer species and the community becomes more complex with larger species.

> **Now test yourself**
>
> 1 Explain why the most diverse community is found just before the climax community is reached.
>
> Answer on p. 125
>
> Tested

Measuring distribution and abundance of organisms

The distribution and abundance of organisms can be measured in a number of ways:

- **Line transects** are used where conditions change and the distribution of organisms also changes. A line transect is a long tape measure or rope laid out on the ground. Observations are made along the transect and every species touching it is recorded, along with an estimate of its abundance. In a long transect, observations may be made at suitable intervals.

- **Belt transects** are similar to line transects, but involve a band rather than a single line. This means that each unit length of the belt transect provides an area in which abundance or percentage cover can be measured accurately rather than estimated.

- **Quadrats** are square frames of a suitable size. They are placed over a random site and examined closely to identify all the plants inside the quadrat. A quantitative sample can be achieved by measuring percentage cover of each species within the quadrat. This can be done using the following methods:
  - point sampling using a **point quadrat** — place a point frame in the quadrat and count the number of examples of each species that touch each point
  - divide the quadrat using string into smaller squares, then estimate how many squares are occupied by each species

> **Revision activity**
>
> Look back at your work from Unit F212 and the ways in which biodiversity was measured.

> **Examiner's tip**
>
> This is an obvious topic for synoptic testing of material in F212.

The role of decomposers

Decomposers are microorganisms that cause decay and recycle minerals. They release enzymes into their surroundings and absorb the small organic molecules produced by digestion. Decomposers are essential to break down all the large organic molecules that have been produced by autotrophs and heterotrophs. This includes:

- uneaten plants and animals that die
- parts of organisms that are shed, such as leaves and exoskeletons
- the remains of plants and animals that have been partly eaten
- undigested material egested as faeces
- compounds that have been excreted

The **decomposition** of these large complex molecules is essential to return the simple inorganic components to the soil so that they can be used again by plants.

Recycling nitrogen

Nitrogen is essential for the production of amino acids and proteins, as well as for DNA. The **recycling of nitrogen** is shown in Figure 12.1.

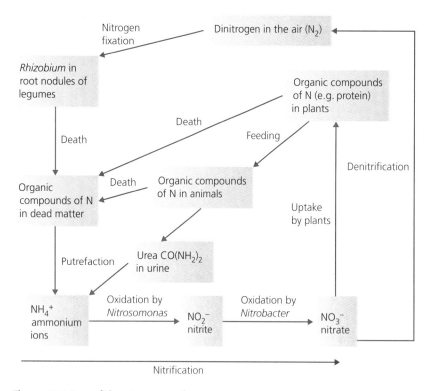

Figure 12.1 Part of the nitrogen cycle

As you can see from the figure, bacteria play an essential role in recycling nitrogen. *Nitrosomonas* and *Nitrobacter* are known as nitrifying bacteria. They are involved in converting complex nitrogen-containing compounds to simple inorganic compounds that can be absorbed by plants. *Nitrosomonas* converts ammonium salts to nitrites, whereas *Nitrobacter* converts nitrites to nitrates. *Rhizobium* is a nitrogen-fixing bacterium that lives free in the soil but also forms an association with plants in the legume family. It lives in special nodules in the roots of legumes and supplies the plant with ammonium salts that are converted to nitrates and used in producing proteins.

Revision activity

Sketch a diagram of the nitrogen cycle, writing notes beside each arrow to explain the process represented by the arrow.

Now test yourself

Tested ☐

3 Explain why farmers in many parts of the world plant crops of beans or grass and clover every few years.

Answer on pp. 125–6

Exam practice

1 (a) (i) What is meant by the term *producer*? [1]

(ii) Explain why a producer is only able to use about 1% of the sun's energy. [4]

(b) The transfer of energy between producers and primary consumers can be as low as 5%. Describe and explain how farmers improve the efficiency of energy transfer between these trophic levels. 1 mark is available for making clear the link between the change made and the effect it has on improving efficiency. [7]

2 (a) Explain the role of decomposers in an ecosystem. [3]

(b) Explain the reasoning behind the following statements:

(i) Sowing clover every fourth year improves productivity. [2]

(ii) Ploughing in the roots and stems of crops after harvest improves the soil. [2]

(iii) Maintaining hedgerows and wild areas on farms reduces damage to crops by pests. [2]

(iv) More people can be fed per hectare of land if they eat a vegetarian diet rather than meat. [2]

(v) Planting native species of trees improves biodiversity more than planting imported species. [2]

Answers and quick quiz 12 online

Online ☐

Examiner's summary

By the end of this chapter you should be able to:
- Define the terms *ecosystem, biotic factor, abiotic factor, producer, consumer, decomposer* and *trophic level*.
- State that ecosystems are dynamic systems.
- Describe how energy is transferred though ecosystems.
- Outline how energy transfers between trophic levels can be measured and discuss the efficiency of energy transfers between trophic levels.
- Explain how human activities can manipulate the flow of energy through ecosystems.
- Describe one example of primary succession resulting in a climax community.
- Describe how the distribution and abundance of organisms can be measured, using line transects, belt transects, quadrats and point quadrats.
- Describe the role of decomposers in the decomposition of organic material and describe how microorganisms recycle nitrogen within ecosystems.

13 Populations and sustainability

Population size

Limiting factors

A population of organisms in a new environment will grow exponentially as long as there are no limitations to growth. The sigmoid curve seen in Figure 10.1 on p. 80 would be followed through the lag and exponential growth phases. However, eventually some factor will limit the population growth and its final size.

The eventual maximum size of a population that can be sustained by its environment is known as the **carrying capacity**. This may be reached by a smooth curve or the population may overshoot and fall back to the final size. **Limiting factors** include the following:

A **limiting factor** is one component of the ecosystem that limits the growth and final size of the population.

- Availability of resources such as food, water, light (for plants), oxygen, space or suitable nest sites. When such resources are scarce, competition occurs.

- Number of **predators** — if there are few predators, the **prey** population can grow. However, a large population of predators feeding on their prey will reduce the prey population size. As they feed, the predator population may grow, and if they overuse the prey population (eating too many) then the predator population will fall as the prey population falls. This can cause a cyclical effect — as an increasing population of predators eats more prey, the prey population falls, followed then by a reduction in the predator population. Examples include the lynx and snowshoe hare.

Revision activity

Draw a mind map with 'population size' at the centre. Describe and explain the effects of all the factors that affect the size of a population.

- Disease — in order to survive, the pathogen causing a disease must be transmitted from host to host. If the host population is at a very low density, the pathogen does not spread. However, at higher density the pathogen can be transmitted easily and will cause disease, reducing the population size.

- Behaviour — certain types of behaviour can affect population size. As population density increases, greater aggression or migration can limit population size (e.g. lemmings). In some species, pregnant females may reabsorb their embryos when the population gets too high (e.g. rabbits).

- Natural disasters.

- Climate and weather.

Interspecific and intraspecific competition

Competition occurs when resources are limited.

Interspecific competition is between individuals of different species. It affects the size of both populations and the distribution of both species. Two species cannot live in complete competition. Where this happens, it is likely that one species will adapt to become a slightly better competitor and will out-compete the other. This may lead to:

- extinction of one species
- greater specialisation of each species to avoid competition, which may involve concentrating on one type of food or using a specific type of nest site
- a change in the distribution of the species, so avoiding competition by living in different areas. This may be because conditions in the two areas are different and give a slight advantage to one species. For example, in most of England the grey squirrel has out-competed the red squirrel. However, the red squirrel still survives and out-competes the grey in coniferous woodland.

Intraspecific competition is between individuals of the same species. As a population grows, resources become more limited and competition increases so that only the best adapted — the best competitors — survive. This reduces the growth and limits the eventual size of the population. If the population falls, competition is reduced and the population rises again. Intraspecific competition takes two forms:

- a dominance hierarchy, in which only the strongest and most dominant individuals survive to pass on their alleles
- scramble competition, in which all individuals have an equal chance of surviving and passing on their alleles. This may lead to evolution as variation between individuals confers a selective advantage on those that are better competitors.

> **Examiner's tip**
>
> Don't forget that competition occurs only once resources are limited and the effect of competition on evolution is important — it allows natural selection to occur.

> **Now test yourself**
>
> 1 Suggest why scramble competition rather than a dominance hierarchy is more likely to lead to natural selection.
>
> Answer on p. 126
>
> Tested

Managing ecosystems

Conservation and preservation

Conservation is a **dynamic process** in which positive action is taken to maintain and enhance biodiversity, such as:

- active breeding programmes
- importing individuals from captive breeding programmes or from other more stable ecosystems to increase population size
- removal of excess predators
- prevention of poaching
- monitoring health and carrying out vaccination to reduce disease
- provision of food to reduce competition
- active **management** to provide suitable diverse habitats
- **reclamation** of land in order to return it to the original habitat and ease competition for space

> **Conservation** is an active process that seeks to increase biodiversity.

Preservation is maintaining or protecting an area in its current state, and is best applied to areas that have not yet been affected by humans. If the area has already been altered, it may not have the suitable environmental conditions to enable survival of the ecosystem. It may allow succession to occur or the ecosystem may become unstable.

> **Preservation** is maintaining or protecting an area in its current state.

Now test yourself   **Tested** ☐

2   Explain why preservation is not enough to maintain species diversity.

Answer on p. 126

Sustainable woodland management   Revised ☐

Suitable management of an ecosystem can achieve a wide range of goals that may, at first, appear to conflict. For example, woodland in a **temperate** country can be managed to produce a **sustainable** harvest while still allowing amenity use and maintaining biodiversity. This involves a number of simple techniques:

> A **sustainable** harvest is one that can be carried out indefinitely without damaging the ecosystem.

● Never clear fell — use selective thinning or selective felling so that the ground is never fully exposed over a large area. Strip felling is an alternative.

● Replace felled trees using fast-growing native species.

● Protect young trees from browsing by rabbits and deer.

● Harvest timber from a part of the woodland each year and rotate the area used to avoid clearing one huge area. This practice also means that a wide range of habitats is maintained as each year new habitats are created by harvesting.

● Carry out coppicing (cutting the plants at ground level leaving the roots so that the soil is not disturbed) — this encourages more growth and produces timber that has a variety of traditional uses or can be used as fuel. The cut branches can be stacked and allowed to decay slowly, which provides a range of habitats for many small birds, mammals and invertebrates.

● Carry out pollarding (cutting all the branches off taller trees at 3–6 m above the ground) — this has the same advantages as coppicing.

● Use standards — these are tall trees left to grow between coppiced plants, which protect the soil. These trees can also be harvested once they are large enough.

● Include rides or fire breaks to protect the crop from fires. These also double as walking trails or cycle tracks.

● Incorporate other land uses in the woodland, such as:
  – paths for walkers or off-road cyclists
  – leisure activities such as camping, orienteering or paint-balling
  – conservation — leave piles of scrub/dead wood for habitats

Typical mistake

Some candidates forget to mention the rotational aspect of harvesting, but this is perhaps the most important thing to mention.

Now test yourself

3   Explain why rotational coppicing is a good woodland management technique.

Answer on p. 126

Tested ☐

Economic, social and ethical reasons

Conservation programmes concentrate on maintaining biodiversity. The reasons for conservation include the following:

- Every species has its own value and a right to survive. As many species are endangered as a result of human activities, we have an ethical responsibility to conserve them.
- Many species have an economic value when harvested as a food source or fuel supply.
- Many species have a future potential as a source of genetic diversity for breeding new characteristics into our crops or as a source of medicinal drugs.
- The provision of natural predators to pests that damage crops.
- The pollination of crops.
- The provision of so-called environmental services, such as maintenance of water quality, the water cycle and soil quality, recycling minerals and the decomposition of waste.
- Ecotourism also brings wealth to some areas.

Typical mistake

This is another example of a topic where some candidates write inappropriate lengthy arguments about the ethics and morals of human activities — at A2 it is most appropriate to stick to the facts.

Revision activity

Draw a mind map with 'conservation' at the centre. Include all the reasons for conservation, grouping the reasons according to whether they are economic, social or ethical.

Human activities in the Galapagos Islands
Revised

Since Charles Darwin's discoveries made the Galapagos Islands famous, they have been severely affected by human activities:

- fishing and whaling, which have upset the marine ecosystem, in particular fishing of marine cucumbers and lobsters
- the introduction of new species, such as:
  - goats which compete with native species for vegetation
  - dogs and cats which chase and eat native species
  - rats and mice which damage the eggs of native species
  - plants such as elephant grass which compete with native vegetation
- tourism
- disturbance due to scientific research and collecting samples
- increasing population, which requires more land for housing and agriculture, produces sewage and waste, and uses more water and energy supplies

Examiner's tip

Remember that questions could link back to conservation in F212 and ask about what conservation techniques could be used to reduce these effects.

Exam practice

1 (a) Explain what is meant by the term *limiting factor*. [2]
  (b) Describe and explain the effect that removing all the predators from an area may have on the ecosystem. [3]
  (c) Predators are a limiting factor. List three other limiting factors on the size of a population. [3]
2 (a) Describe one example of interspecific competition. [3]
  (b) Explain how interspecific competition could lead to the extinction of one species. [3]
  (c) Explain why conservationists are concerned about the introduction of goats to the Galapagos Islands. [3]

Answers and quick quiz 13 online

Online

Examiner's summary

By the end of this chapter you should be able to:

✔ Explain the significance of limiting factors in determining the final size of a population and explain the meaning of the term *carrying capacity*.

✔ Describe predator–prey relationships and their possible effects on population sizes.

✔ Explain the terms *interspecific competition* and *intraspecific competition*.

✔ Distinguish between the terms *conservation* and *preservation*.

✔ Explain how the management of an ecosystem can provide resources in a sustainable way, with reference to timber production in a temperate country.

✔ Explain that conservation is a dynamic process involving management and reclamation.

✔ Discuss the economic, social and ethical reasons for conservation of biological resources.

✔ Outline the effects of human activities on the animal and plant populations in the Galapagos Islands.

14 Plant responses

Responding to environmental changes

Examiner's tip

This examination paper includes synoptic marks. These test your:
- understanding of the principles behind different processes
- ability to make links back to other parts of the specification

The obvious links here are:
- cell signalling from F211
- communication and hormones from F214

Why do plants need to respond?
Revised

All organisms need to respond to changes in their internal or external environment. The changes are stimuli and the responses made act to increase the survival chances of the organism. Plants respond to avoid:

- **predation**
- **abiotic stress**, which includes insufficient light or water

Predation means the preying on one organism by another.

Abiotic stress is the negative impact of non-living factors on the living organisms in a specific environment.

Tropisms
Revised

Plant responses are called **tropisms** and they include:

- phototropism (the response to light) — shoots show positive phototropism and grow towards light

- geotropism (the response to gravity) — shoots show negative geotropism and grow away from gravity. This ensures they grow up into the light. Roots show positive geotropism and grow towards gravity. This ensures they grow down into the soil.

- chemotropism (the response to certain chemicals) — the pollen tube grows towards the ovum during reproduction

- thigmotropism (the response to touch) — some plants grow so that their stem or branch winds around a support

A **tropism** is a directional growth movement.

Now test yourself

1 Explain why a plant responding to touch may enhance its chances of survival.

Answer on p. 126

Tested

Plant hormones
Revised

Plant responses are coordinated by growth regulators, which are often called **hormones**. These are similar to animal hormones in a number of ways:

- They are produced only in particular tissues.
- They move around the whole body.

- They are specific and act on particular target tissues.
- The target tissue cells have specific complementary receptors in their cell surface membranes.

However, plant hormones are different from animal hormones in a number of ways:

- Plants do not have endocrine glands — their hormones are released from a variety of tissues.
- The hormones are not secreted into the transport system.
- They move by diffusion, by active transport and also in the phloem.
- They may act at very low concentrations, but their precise effect may depend on the concentration.
- They may have a different effect on different tissues.
- Two or more hormones may work together to amplify their effects (synergy) or to cancel out their effects (antagonism).

Phototropism
Revised

Phototropism is the growth response to light. The response to light is coordinated by the hormone **auxin**, which is produced in the tip of a growing shoot. As it diffuses down the shoot, it increases growth by promoting further **elongation** of the cells. The mechanism involves pumping hydrogen ions into the cell wall so that the change in pH softens the cell wall, allowing extra stretching. However, auxin distribution is affected by light as it accumulates in higher concentrations on the shaded side of the shoot. The mechanism of this redistribution is unclear, but it involves the activity of enzymes and auxin receptors in the plasma membranes. This causes the cells on the shaded side to elongate more than on the brightly lit side, resulting in the shoot growing slightly longer on the shaded side and bending towards the light.

> **Examiner's tip**
>
> Remember that the bending of the shoot is caused by extra growth or elongation of the cells on the shaded side of the stem.

> **Now test yourself**
>
> 2 Explain why a plant responding to light enhances its chances of survival.
>
> Answer on p. 126
>
> Tested

Experimental evidence
Revised

Apical dominance refers to the fact that auxin is produced at the tip of the shoot (the apex) and that it inhibits the growth of lateral (side) buds that will form side shoots. Therefore, when the tip is intact the plant tends to grow upwards rather than out to the sides. The evidence for this includes the following:

- The removal of the apical bud allows the lateral buds to grow.
- Auxin or synthetic auxin placed on the cut tip continues to inhibit the growth of side shoots.
- Using a chemical that inhibits auxin transport allows lateral bud growth.
- Later investigations have revealed that the link is not direct. Auxin promotes the production of abscisic acid, which inhibits the lateral bud growth. When auxin levels drop, so does the concentration of abscisic acid so the bud can grow. Its growth is further stimulated by the accumulation of cytokinins that collect where auxin is most concentrated, so when the apical bud is removed the cytokinins accumulate in the lateral buds.

> **Apical dominance** refers to the inhibition of lateral shoots by the apical bud.

> **Examiner's tip**
>
> In terms of factual recall, the plant responses topic is quite light. However, candidates are expected to be able to answer questions based on experimental evidence.

Elongation of the stem is stimulated by **gibberellins**. The evidence for this includes the following points:

- Certain fungi have been shown to increase seedling growth, and chemicals isolated from these fungi have the same effect.
- Applying gibberellins externally has the same effect.

Leaf loss (abscission)　　　　　　　　　Revised

Deciduous plants shed their leaves each year. There is a region of cells at the end of the leaf stalk called the abscission zone, which responds to a number of plant hormones:

- Auxin produced in the leaf inhibits abscission by making the cells insensitive to ethene.
- Cytokinins enter the leaf and stop the leaves ageing.
- When the concentration of cytokinins in the leaf falls, the leaf ages and turns brown. This is called senescence and stops the production of auxin in the leaf.
- The cells in the abscission zone become sensitive to ethene, which stimulates the production of the enzyme cellulase in the abscission zone.
- Cellulase digests the cell walls in the abscission zone and the leaf falls.

Commercial uses of plant hormones　　　　　　Revised

Plant hormones are used widely in a variety of industries, as shown in Table 14.1.

Table 14.1 Commercial uses of plant hormones

Hormone	Commercial uses
Auxins	To promote root growth in cuttings
	To produce seedless fruit
	As a selective weed killer
Cytokinins	To delay leaf senescence to avoid vegetables discolouring, which increases shelf life
	To promote bud and shoot growth during tissue culture
	To promote the growth of lateral buds
Ethene	To speed up fruit ripening and promote fruit drop
	To promote the growth of lateral branches
Gibberellins	To delay fruit senescence and drop to make harvesting more efficient
	To improve the shape and size of fruit
	To activate enzymes in stored barley to produce malt for brewing
	To speed up seed production during breeding programmes

Exam practice

1 A student investigated the effects of auxin and gibberellins on plant growth. She took ten shoots growing in a suitable medium. Five were coated with gibberellin paste and five were coated with auxin paste. She measured the length of the lateral shoots each day for 16 days. Her results are shown in the following graph.

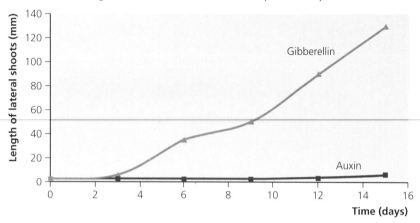

(a) Suggest a suitable control for this experiment. [2]

(b) Suggest why the student used five shoots in each sample. [3]

(c) After 15 days the shoots with auxin had grown to 6 mm and those with gibberellins had grown 130 mm. The student concluded that gibberellins have a greater effect on the growth of the lateral shoots than auxin. Calculate the percentage increase in growth at 15 days of shoots with gibberellins applied compared to those with auxin applied. [2]

(d) Suggest why the student's conclusion may not be accurate. [2]

2 A student cut 5 mm sections of young growing stem from several similar plants of the same species. The 5 mm sections were placed in Petri dishes containing auxin solutions of different concentrations. After 12 hours the sections were removed from the Petri dishes and measured. The following graph shows the mean increase in the length of the sections in each dish, plotted against the concentration of auxin.

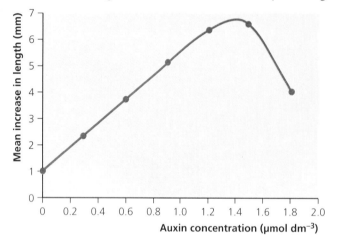

(a) (i) Describe the relationship between the concentration of auxin and the mean increase in length of the stem sections. [3]

   (ii) List three variables that should have been controlled in the investigation. [3]

(b) Suggest two ways in which auxin might have caused the change in growth of the stems. [2]

(c) Describe three commercial uses of plant hormones. [3]

Answers and quick quiz 14 online

Online

Examiner's summary

By the end of this chapter you should be able to:

✔ Explain why plants need to respond to their environment in terms of the need to avoid predation and abiotic stress.

✔ Define the term *tropism*.

✔ Explain how plant responses to environmental changes are coordinated by hormones, with reference to responding to changes in light direction.

✔ Evaluate the experimental evidence for the role of auxins in the control of apical dominance and gibberellin in the control of stem elongation.

✔ Outline the role of hormones in leaf loss in deciduous plants.

✔ Describe how plant hormones are used commercially.

15 Animal responses

Responding to environmental changes

Examiner's tip

This examination paper includes synoptic marks. These test your:
- understanding of the principles behind different processes
- ability to make links back to other parts of the specification

The obvious links here are:
- cell signalling from F211
- communication, nerves, hormones and homeostasis from F214

Why do animals need to respond? — Revised

All organisms need to respond to changes in their internal or external environment. The changes are **stimuli** and the **responses** made act to increase the survival chances of the organism. Animals respond to:

- avoid predation
- avoid abiotic stress, which includes extremes of temperature, insufficient water and irritants
- find food
- increase the chances of reproduction

Typical mistake

Candidates often forget to consider the survival advantage of responses.

Stimuli (singular: stimulus) are changes in the environment that bring about a response.

Responses are changes in behaviour that result from a stimulus.

Revision activity

Draw a mind map with 'response' at the centre. Include both plants and animals and note how and why they respond.

The nervous system — Revised

The **nervous system** must detect environmental changes and bring about suitable responses to those changes. There are receptors all over the body to detect a variety of environmental stimuli, both internal and external. The nerves transmit the information and the brain receives this information from the receptors. The brain puts all the information together in association centres and coordinates suitable responses via the motor areas. The nervous system can be divided into two regions:

- the **peripheral nervous system (PNS)**, which consists of all the receptors, the sensory neurones and the motor neurones, and includes the autonomic nervous system and the somatic nervous system.

- the **central nervous system (CNS)**, which consists of the brain and spinal cord

The autonomic nervous system

Revised

The **autonomic nervous system (ANS)** is a part of the PNS that coordinates unconscious responses to do with maintaining the internal environment. Receptors detect changes in the internal environment and association centres in the brain coordinate a response by sending action potentials down autonomic pathways to the effectors. All the cells are motor neurones and most are unmyelinated.

> The **autonomic nervous system (ANS)** is a series of motor neurones that coordinate the unconscious responses involved in homeostasis.

The pathway that carries action potentials from the CNS to the effector consists of at least two neurones with synapses inside swellings called ganglia (singular: ganglion). The ANS comprises two antagonistic sets of nerves: the parasympathetic nerves and the sympathetic nerves (Table 15.1).

Table 15.1 Comparing the parasympathetic and sympathetic systems

System	Parasympathetic system	Sympathetic system
Organisation	Only a few neurones leading out of the CNS (including the vagus nerve), which divide up and lead to different organs	Many neurones leading out of the CNS
Position of ganglia	In the effector tissue	Just outside the CNS
Length of pre-ganglionic neurones	Long	Short
Length of post-ganglionic neurones	Short	Long
Neurotransmitter	Acetylcholine	Noradrenaline
Role	Decreases activity — to conserve energy	Increases activity — to prepare for activity
When active	Most active in sleep or relaxation	Most active at times of stress
Effects	Decreases heart rate	Increases heart rate
	Constricts pupils	Dilates pupils
	Reduces ventilation rate	Increases ventilation rate

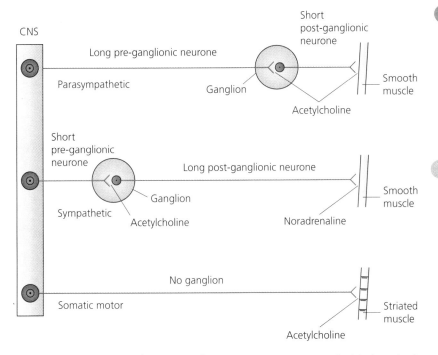

Figure 15.1 Neurones in the autonomic nervous system compared with those in the somatic motor system

Revision activity

Draw an outline of a human body and inside add a simple version of the nervous system, including the autonomic system. Annotate your diagram to show the function of each part of the nervous system.

Now test yourself

1 Explain why there are two parts to the autonomic nervous system.

Answer on p. 126

Tested

Structure and functions of the human brain

The brain is divided into four sections: the cerebrum, cerebellum, medulla oblongata and hypothalamus (Table 15.2).

Table 15.2 The structure and functions of the major parts of the brain

Area of brain	Structure and function
Cerebrum	The cerebrum is the largest part of the human brain
	It is divided into two hemispheres joined by the corpus callosum
	The cerebral cortex (the outer part of the cerebrum) is highly folded. It deals with 'higher functions' such as conscious thought, overriding some reflexes, intelligence, reasoning, judgement and memory
	The cerebral cortex is divided into:
	● sensory areas that receive impulses from the sensory neurones
	● motor areas that send impulses out to the effectors
	● association areas that link information together and coordinate the appropriate response
Cerebellum	The cerebellum coordinates fine control of muscular movement, walking, running using tools, etc.
	It coordinates balance and body position
	It receives information from the various senses (retina, balance organs in the ear, muscle spindles, joints) and sends impulses to the motor areas
Medulla oblongata	The medulla oblongata controls the involuntary processes such as heart rate and breathing
	It contains specialised centres such as the cardiac centre and the respiratory centre, which receive information from internal receptors and modify the heart rate and breathing rate respectively
Hypothalamus	The hypothalamus controls the homeostatic mechanisms
	It receives information from a variety of receptors, mostly internal
	It monitors the blood and regulates factors such as body temperature and blood water potential
	It controls the endocrine system via the pituitary gland

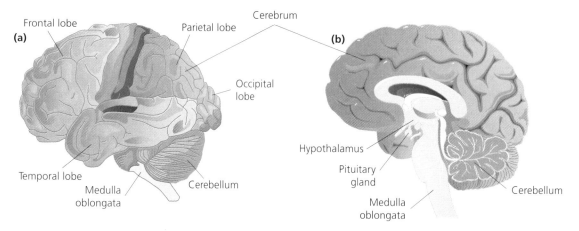

Figure 15.2 The human brain showing the major parts

The pathway of a nerve impulse from stimulus to response involves the CNS:

stimulus → receptor → sensory neurone → CNS → motor neurone → effector → response

Figure 15.3 shows what happens when a student writes down a word that the teacher has written on the board. This pathway is under conscious control and forms part of the somatic nervous system.

Revision activity

Draw a sketch of the human brain and annotate it to show the functions of each part.

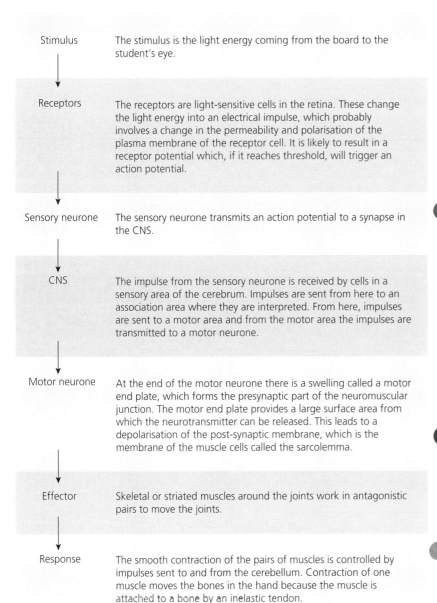

Stimulus	The stimulus is the light energy coming from the board to the student's eye.
Receptors	The receptors are light-sensitive cells in the retina. These change the light energy into an electrical impulse, which probably involves a change in the permeability and polarisation of the plasma membrane of the receptor cell. It is likely to result in a receptor potential which, if it reaches threshold, will trigger an action potential.
Sensory neurone	The sensory neurone transmits an action potential to a synapse in the CNS.
CNS	The impulse from the sensory neurone is received by cells in a sensory area of the cerebrum. Impulses are sent from here to an association area where they are interpreted. From here, impulses are sent to a motor area and from the motor area the impulses are transmitted to a motor neurone.
Motor neurone	At the end of the motor neurone there is a swelling called a motor end plate, which forms the presynaptic part of the neuromuscular junction. The motor end plate provides a large surface area from which the neurotransmitter can be released. This leads to a depolarisation of the post-synaptic membrane, which is the membrane of the muscle cells called the sarcolemma.
Effector	Skeletal or striated muscles around the joints work in antagonistic pairs to move the joints.
Response	The smooth contraction of the pairs of muscles is controlled by impulses sent to and from the cerebellum. Contraction of one muscle moves the bones in the hand because the muscle is attached to a bone by an inelastic tendon.

Figure 15.3 Flow chart showing the pathway of the nerve impulse involved in copying a word from the board

Examiner's tip

You are expected to apply relevant knowledge about each step from stimulus to response. Remember that you have studied the structure of neurones and synapses (F214), the structure of cell membranes (F211) and the action of synapses and transmission of action potentials (F214). Even the action of enzymes in the synaptic cleft may bring in information from F212.

Typical mistake

Candidates often fail to distinguish between the stimulus and the receptor and between the effector and the response.

Revision activity

Draw a flow chart similar to Figure 15.3, using catching a ball as another example of conscious behaviour.

Movement and muscular coordination

The action of muscles at joints
Revised

Movement requires the coordinated action of voluntary muscles (skeletal muscles). Skeletal muscles can contract (get shorter), but they cannot elongate again without the action of an **antagonistic** muscle. Therefore, skeletal muscles work in antagonistic pairs. Movement is achieved by coordinated action of the antagonistic muscles — when one muscle contracts to get shorter, the other relaxes and is pulled out or extended.

The muscles are stimulated by motor neurones leading from the motor area of the brain. One motor neurone is attached to one or more muscle fibres, called a motor unit. The neurone is attached to the muscle fibre by a motor end plate or neuromuscular junction. This works just like a synapse to stimulate the muscle fibre membrane.

Antagonistic means that the muscles oppose each other.

A typical synovial joint: the elbow

Revised

The arm contains three main bones: the humerus, radius and ulna, which join at the elbow. The elbow is a hinge joint. The biceps muscle is a flexor muscle that bends the joint when it contracts. The triceps is an extensor that straightens the joint when it contracts. The biceps and triceps are antagonistic — they work against one another (Figure 15.4).

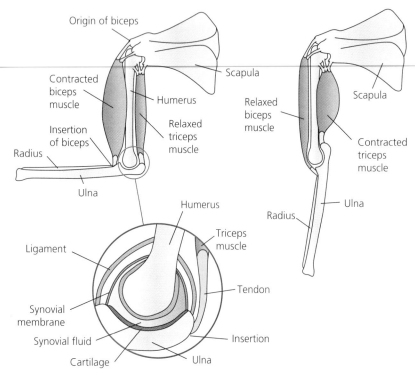

Figure 15.4 Antagonistic action of the muscles that move the forearm about the hinge joint at the elbow

Now test yourself

2 Explain why muscles are always found in antagonistic pairs.

Answer on p. 126

Tested

Skeletal muscle

Revised

Skeletal muscle is banded or striated, so it is also called striated muscle. The **striations** are produced by the arrangement of the actin and myosin protein filaments inside the muscle fibres. The dark band (A band) is made up of thicker myosin filaments that are held together by the M line. The lighter band (I band) is made up of thinner actin filaments that are held together by the Z line. The distance between two Z lines is called a sarcomere and this is the active unit of a muscle. The actin and myosin filaments overlap.

> **Striations** are banding patterns created by the protein filaments in the fibre.

Each fibre contains many nuclei — it is multinucleate. The fibres are surrounded by a membrane known as the sarcolemma, which is equivalent to the plasma membrane of cells. The cytoplasm is called sarcoplasm and is specialised as it contains many mitochondria to supply energy to the **contraction** process. The endoplasmic reticulum is also specialised to form sarcoplasmic reticulum, which is involved in controlling contractions.

Contraction is brought about by the actin and myosin filaments sliding past one another. This moves the Z lines closer together so the sarcomeres get shorter, as does the whole fibre. During contraction, the banding pattern changes:

- The I band gets narrower.
- The H zone (the region of the A band where there is no overlap with the actin) gets shorter.

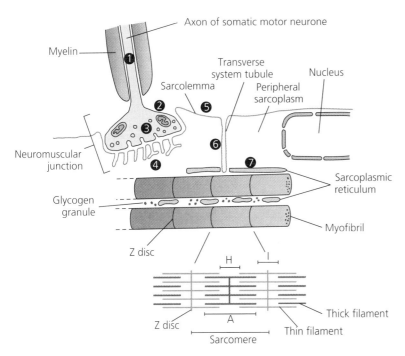

Figure 15.5 The events that occur in a striated muscle fibre following stimulation by a motor neurone

1 Action potentials travel down the motor neurone.

2 They stimulate calcium ions to enter the presynaptic swelling.

3 The vesicles of neurotransmitter fuse with the membrane releasing acetylcholine into the cleft.

4 Acetylcholine binds to receptors, causing an inflow of sodium ions that depolarise the sarcolemma (an end-plate potential).

5 If above threshold, action potentials are transmitted along the sarcolemma.

6 Action potentials are also transmitted down into the transverse system (T-system) tubules. These membranes have the same channel proteins and pumps found in neurones.

7 The impulse is coupled to the contraction mechanism. Depolarisation of T-system tubules causes calcium channels in the sarcoplasmic reticulum to open and release calcium ions into the sarcoplasm. These act as second messengers to stimulate movement of the muscle myofibrils.

The myosin filaments have specialised regions called heads. These **myosin heads** can bridge the gap between the actin and the myosin. When the myosin head attaches to a special binding site on the actin, it changes shape and pulls the actin past the myosin, making the filaments slide (the **sliding filament model**).

> **Now test yourself**
>
> 3 Explain why the I band and the H zone get narrower during contraction.
>
> Answer on p. 126
>
> Tested

> A **myosin head** is a part of the myosin filament that can move and attach to the actin filament.

The myosin head is an ATPase. When the actin and myosin are locked together, **ATP** joins the myosin head. It is broken down, releasing energy that separates the actin and myosin and moves the myosin head back to its starting position. When the myosin rebinds to the actin and slides the actin forward, the ADP and P$_i$ are released (Figure 15.6). ATP supplies are maintained by a reaction involving creatine phosphate (CP). Muscle tissue contains CP, which can transfer phosphate to ADP to make ATP. Muscle tissue also contains glycogen, which can be used anaerobically or aerobically in respiration.

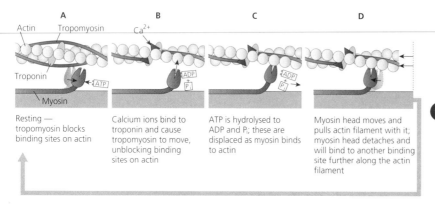

Resting — tropomyosin blocks binding sites on actin	Calcium ions bind to troponin and cause tropomyosin to move, unblocking binding sites on actin
ATP is hydrolysed to ADP and P$_i$; these are displaced as myosin binds to actin	Myosin head moves and pulls actin filament with it; myosin head detaches and will bind to another binding site further along the actin filament

Typical mistake

Remember that ATP is broken down to release energy for the recovery stroke rather than for the power stroke.

Figure 15.6 The sliding filament model: changes that occur in a myofibril during contraction

The role of calcium ions

Revised

Calcium ions are stored in the sarcoplasmic reticulum. When the muscle is stimulated by a motor neurone the action potential travels along the sarolemma and down the T-system tubules. This action potential causes the release of calcium ions from the sarcoplasmic reticulum. The calcium ions bind to the protein troponin which moves the tropomyosin aside. This exposes binding sites on the actin which the myosin heads can bind to and cause contraction.

Synapses and neuromuscular junctions

Revised

Synapses and **neuromuscular junctions** are similar in structure and function (Table 15.3).

Table 15.3 Comparing synapses and neuromuscular junctions

Synapse	Neuromuscular junction
Activated by action potentials	Activated by action potentials
Release the transmitter acetylcholine by exocytosis into the cleft	Release the transmitter acetylcholine by exocytosis into the cleft
Diffusion of transmitter across the cleft	Diffusion of transmitter across the cleft
Receptor sites on the post-synaptic membrane open sodium channels to produce an action potential in the membrane	Receptor sites on the post-junction membrane open sodium channels to produce an action potential in the membrane
Action potential in post-synaptic cell	Muscle contracts

Muscle types

Revised

There are three types of muscle in the body: **voluntary muscle**, **involuntary muscle** and **cardiac muscle**. Each is different in structure and function (Table 15.4).

Table 15.4 Comparing voluntary, involuntary and cardiac muscle

Voluntary muscle	Involuntary muscle	Cardiac muscle
Striated, multinucleate, organised in parallel bundles of myofibrils; fibres all parallel	Smooth muscle, consists of individual cells	Organised as single cells joined by intercalated discs; parallel myofibrils similar to striated muscle; cross bridges present between fibres
Skeletal muscle attached to bones by tendons	Found in walls of blood vessels, digestive system and airways	Found only in the heart
Supplied by nerves from the peripheral somatic nervous system	Supplied by nerves from the autonomic nervous system	Supplied by nerves from the autonomic nervous system

Now test yourself

Tested

4 Suggest why cardiac muscle has cross bridges between the fibres.

Answers on p. 126

Examiner's tip

When asked to compare or contrast in an exam question, you can draw a table in the answer space. This is a quick and easy way to make comparisons.

The 'fight or flight' response

Revised

It is important to remember that many responses are coordinated by both the nervous and the **endocrine system**. One example is the **'fight or flight'** response seen in mammals. In a dangerous or stressful situation, the body is prepared for activity. This involves a series of changes:

1 The sensory receptors detect the environmental changes.

2 Sensory neurones carry action potentials to the CNS.

3 Spinal or cranial reflex actions may bring about very rapid responses.

4 Impulses are also conducted to the cerebrum, which uses the association centres to make decisions about how to respond.

5 Impulses are sent down the somatic motor neurones to the skeletal muscles to bring about coordinated voluntary movement.

6 The hypothalamus is stimulated and sends impulses down the sympathetic part of the autonomic nervous system to bring about a range of changes:

● The cardiac accelerator nerve carries impulses to the heart, which increase the heart rate and stroke volume.

● Blood pressure is increased.

● Breathing rate and depth may rise.

● Blood vessels to the gut and skin constrict.

● Blood vessels to the muscles dilate.

● The adrenal glands are stimulated to release adrenaline, which stimulates the liver to release glucose and maintains the other effects of the sympathetic nervous system.

Now test yourself

5 Suggest why the 'fight or flight' response is coordinated by both the nervous and the endocrine systems.

Answer on p. 126

Tested

Exam practice

1 **(a)** Complete the following table to describe the functions of parts of the human brain. [4]

Area of brain	Function
Cerebrum	
	Control of heart rate
	Thermoregulation
Cerebellum	

(b) Complete the following diagram to show the organisation of the nervous system. [6]

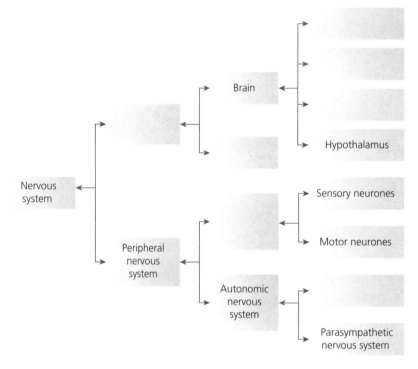

2 **(a)** The cerebellum in the brain is concerned with the control and coordination of movement and posture. Suggest why the cerebellum of a chimpanzee is relatively larger than that of a cow. [2]

(b) Discuss the role of the autonomic nervous system in controlling heart rate. [4]

3 **(a)** Describe the mechanism of muscular contraction. [6]

(b) State three differences between cardiac muscle and skeletal (striated) muscle. [3]

Answers and quick quiz 15 online

Online

Examiner's summary

By the end of this chapter you should be able to:

- Discuss why animals need to respond to their environment.
- Outline the organisation of the nervous system in humans.
- Outline the organisation and roles of the autonomic nervous system.
- Describe the gross structure of the human brain and outline the functions of the cerebrum, cerebellum, medulla oblongata and hypothalamus.
- Describe the role of the brain and nervous system in the coordination of muscular movement.
- Describe how coordinated movement requires the action of skeletal muscles about joints, with reference to the movement of the elbow joint.

- Explain the sliding filament model of muscular contraction.
- Outline the role of ATP in muscular contraction, and how the supply of ATP is maintained in muscles.
- Compare the action of synapses and neuromuscular junctions.
- Outline the structural and functional differences between voluntary, involuntary and cardiac muscle.
- State that responses to environmental stimuli are coordinated by the nervous and endocrine systems.
- Explain how, in mammals, the 'fight or flight' response is coordinated by the nervous and endocrine systems.

16 Animal behaviour

Innate and learned behaviours

Innate behaviour Revised

The features of **innate behaviour** include the following:

- It is genetically determined and passed on via reproduction.
- All members of the species show this behaviour.
- It is stereotyped — there is a fixed action pattern.
- It tends to be inflexible — it does not change as a result of changing stimuli around the animal.
- It is unintelligent — the animal has no sense of the purpose of the behaviour.

Innate behaviour always has a survival value. This may be survival of the individual or survival of the species through ensuring reproduction takes place. There are three types of innate behaviour, as shown in Table 16.1.

Table 16.1 Types of innate behaviour

Type of innate behaviour	Example	Survival value
Reflex — usually an **escape** response that takes the body or part of the body out of harm's way very quickly	The escape reflex of earthworms when they are touched or the rapid movement of a squid	To avoid predation (so it does not get eaten)
Taxis — a directional response. The animal moves towards or away from a stimulus	A shark is attracted to a source of blood in the water, so it swims up the concentration gradient	To find more food or to get away from an unpleasant stimulus
Kinesis — this is an increased movement that has no defined direction. The animal moves more quickly in response to an unpleasant stimulus and slows down when that stimulus is no longer present	Woodlice move about more quickly in the light or in dry conditions	To spend more time in the right conditions, where there are no unpleasant stimuli

Examiner's tip

This examination paper includes synoptic marks. These test your:
- understanding of the principles behind different processes
- ability to make links back to other parts of the specification

The obvious links here are:
- cell signalling from F211
- communication, nerves and hormones from F214

Innate behaviour is present at birth and is not taught or learned.

Taxis refers to a directional response.

Kinesis is increased movement in no specific direction.

Typical mistake

Some candidates confuse taxis and kinesis. If you remember that a taxi will take you to a particular place, this is directional.

Now test yourself

1 Explain why maggots move more quickly in bright light.

Answer on p. 126

Tested

Learned behaviour Revised

The following Learned behaviour develops as a result of experience. The features of learned behaviour include the following:

- It is not present at birth.
- Not all members of the species show the behaviour.
- The behaviour can be adapted according to experience.
- It can be passed on to other members of the species by observation.

Learned behaviour is behaviour that is altered by experience.

Learned behaviour usually has a survival value. The types of learned behaviour include those given in Table 16.2.

Table 16.2 Types of learned behaviour

Type of behaviour	Example
Habituation — the animal learns to ignore a repeated stimulus that causes no harm	Birds learn that a scarecrow is no risk People become unaware of a continuous sound or constant smell
Imprinting — young animals associate with another organism	Lorenz showed that young geese (goslings) recognise their own kind because they imprint on the first thing they see that moves — this is usually their parent
Classical conditioning — an animal learns to associate two unrelated stimuli through repetition	Pavlov trained dogs to salivate at the sound of a bell. The salivation is a conditioned response to an inappropriate stimulus
Operant conditioning (or trial-and-error learning) — animals can be trained to perform an act to receive a reward. The reward is a reinforcement	Dogs learn to sit or shake hands if given a reward such as a friendly pat
Latent learning (also known as exploratory learning) — an animal that can explore its surroundings learns the surroundings and may be able to make use of that knowledge to escape predators or find food	Rats and mice can be trained to run through a maze. If they have had previous experience of the maze, they learn it much more quickly
Insight learning — use of previous experience to solve a problem	A chimp trying to reach a banana that is out of reach may stack boxes to climb higher to reach the food

Examiner's tip

This topic has the potential to generate a lot of psychobabble from students, but much of the required knowledge can be reduced to a series of bullet points.

Revision activity

Draw a mind map with 'behaviour' at the centre. Include examples of innate and learned behaviour and explain their survival value.

Now test yourself

2 Explain why a dog salivating at the sound of a bell is not an innate response.

Answer on p. 126

Tested

Social behaviour in primates

Revised

Most primates are social animals living in family groups. Typically, the group has a hierarchical structure in which different individuals have different status and roles. There are many advantages to living in a social group:

- Hunting as a group enables successful hunts for larger animals as food.
- Many individuals searching for new sources of food can cover a wider area.
- Food sources such as a tree in fruit can be protected.
- Many eyes watching for predators will be more likely to spot one coming or individuals can take it in turns to act as a lookout, allowing others to feed in peace.
- Many individuals together can protect each other and fight off a predator.
- Social grooming can help to rid the family of parasites.
- Care of the young can be shared between a number of closely related individuals.
- The young members of the group can learn through observation and play.

Examiner's tip

The specification asks you to use one example to describe the advantages of social behaviour. You do need to learn just one example, not a wide variety.

Understanding human behaviour

Dopamine and DRD4 — Revised ☐

Dopamine is a chemical found in the brain. It is a neurotransmitter and a hormone that is involved in the release of adrenaline and noradrenaline. There are at least five different **dopamine receptors** (called DRD1 to DRD5) and each one mediates a different response. Therefore, dopamine can affect a wide range of activities. Depending on which receptor is activated, dopamine affects behaviour in a different way. The range of responses includes:

- increasing general arousal
- decreasing inhibition
- controlling motivation
- controlling learning

> A **dopamine receptor** is a membrane-bound glycoprotein.

In addition, it is clear that the concentration of dopamine is important. Too little can lead to diseases such as Parkinson's disease and too much can lead to diseases such as schizophrenia. Studies involving synthetic dopamine and dopamine blockers have revealed that addictive behaviour and risk-taking are affected by dopamine levels. A number of antipsychotic drugs are used to block dopamine receptors to help treat these conditions. Attention deficit hyperactivity disorder (ADHD) can be treated with Ritalin, which affects the dopamine levels in the brain.

Each dopamine receptor is coded for by a separate gene. However, natural variation between individuals has led to the existence of at least 50 different alleles for the gene for DRD4. These variations produce receptors that are more or less sensitive to dopamine. Therefore, people with a particular combination of receptors may display behaviour that is not typical or may show certain attributes that other people do not possess. Behaviour affected includes:

Typical mistake

Many candidates confuse the gene and its product — the gene or allele codes for the structure of a protein. That protein is the product which forms the receptor.

- **addictive behaviour** — some people are more likely to display addictive behaviour than others. This may be because of the levels of adrenaline released when behaving in a certain way, such as gambling.
- **risk-taking** — some people are more likely to take risks than others. Again, this could be to do with the levels of adrenaline released when taking a risk: people get a buzz from the risk.

Revision activity

Sketch a synapse between two neurones and annotate the diagram with notes to explain how dopamine acts as a neurotransmitter and how more sensitive receptors may respond to cause the release of more adrenaline.

Examiner's tip

This part of the specification refers to a gene that codes for a receptor protein. In order to fully understand or explain what is going on, it is likely that you will need to draw on knowledge from F212 (protein and DNA structure), the gene control topic in F215 and receptors from F211 and F214.

Now test yourself — Tested ☐

3 Explain how a dopamine blocker could be used to treat compulsive gambling.

Answer on p. 126

Exam practice

1 Reflexes are innate, stereotyped responses to stimuli. However, they can be conditioned.

   (a) Explain the meaning of the terms *innate, stereotyped* and *conditioned.*   [3]

   (b) Describe one example of a reflex response to a named stimulus.   [2]

   (c) A puzzle box is a piece of apparatus used to investigate learning in animals. It presents the animal with a small task and the animal is rewarded if it performs that task correctly. In an investigation, a scientist placed a cat in a box. The box included a loop of string that must be pulled in order to open the box and allow the cat to escape. The scientist placed the cat in the box and recorded the length of time it remained there. The experiment was then repeated several times with the same cat. The results are shown in the following graph.

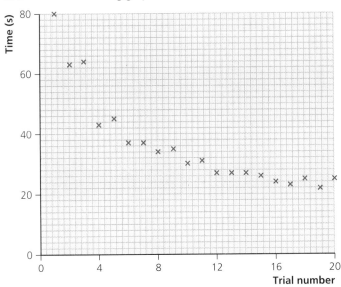

    (i) Which type of learning is being investigated?   [1]

    (ii) Describe the data shown in the graph.   [3]

    (iii) Explain the pattern of data shown in the graph.   [3]

2 Describe, using examples, the advantages of living in family groups.   [7]

Answers and quick quiz 16 online

Online

Examiner's summary

By the end of this chapter you should be able to:

✔ Explain the advantages to organisms of innate behaviour.

✔ Describe escape reflexes, taxes and kineses as examples of genetically determined innate behaviours.

✔ Explain the meaning of the term *learned behaviour.*

✔ Describe habituation, imprinting, classical and operant conditioning, latent and insight learning as examples of learned behaviours.

✔ Describe, using one example, the advantages of social behaviour in primates.

✔ Discuss how the links between a range of human behaviours and the dopamine receptor DRD4 may contribute to the understanding of human behaviour.

Now test yourself answers

1 Communication

1 Positive feedback is usually harmful. If a small change in a parameter brings about an increased change, the system will be unstable. This leads to wild changes in the parameters and enzymes will not be able to function at their optimum capacity. If positive feedback is used in a living system, it can help to magnify a response quickly. However, it must only work until that response has been magnified to an appropriate level. For example, the process of blood clotting involves a cascade effect where a few platelets releasing chemicals cause more platelets to become involved. This produces a clot quickly to reduce blood loss at a wound. However, if this cascade were to continue unchecked, all the blood in the circulatory system would eventually be clotted.

2 Temperature — if temperature drops too low, molecules move more slowly, there will be fewer collisions between molecules and reaction rates go down. If temperature rises too high, the structure of proteins is altered due to bonds within the molecule being broken. The tertiary structure changes and alters the shape of the active site so that substrate molecules no longer fit. The enzyme is denatured and no longer functions.

Water potential of blood — if the water potential of the blood changes, this affects the water in the blood cells and tissue fluid. If blood water potential rises too high, water enters the cells by osmosis and the cells may burst. If blood water potential falls too low, water leaves the blood cells by osmosis and the concentration of all the dissolved substances inside the cells changes. The water potential of the blood affects the formation of tissue fluid and the constituents of the tissue fluid. This will in turn affect the cells surrounded by the tissue fluid.

pH — altering the pH of the blood has a direct effect on protein structure. Extreme pH (high or low) causes changes to the bonding in the active site of enzymes and reduces their ability to function.

Blood glucose concentration — if blood glucose concentration falls too low, insufficient energy is supplied to the body cells, they are not able to respire as much and muscles get tired quickly. Too much glucose in the blood can affect the blood pressure and water potential. Glucose is lost in the urine.

Blood pressure — if blood pressure drops too low, insufficient blood reaches the extremities and tissues may be short of oxygen and glucose supplies. If blood pressure rises too high, this can cause damage to blood vessels and lead to atherosclerosis. It may also damage the kidney or cause a stroke.

Blood concentration of ions such as Na^+, K^+ and Ca^{2+} — the concentrations of these ions affect the water potential of the blood. If the concentration is too high, the water potential drops and blood pressure rises. If the concentration is too low, the water potential rises and blood pressure falls. There may also be an insufficient supply of certain ions to the tissues that require them.

3 Heat can be gained from the environment. If the body is making too much heat, the only way to lose that heat is to the environment. As the skin is the organ that has contact with the environment, it is the surface through which heat can be gained or lost.

2 Nerves

1 Motor neurones have a cell body in the central nervous system. The axon must carry the action potential out to the effector, which may be a muscle at the bottom of the leg. Sensory neurones have their cell body close to the central nervous system and the axon only has to carry the action potential into the central nervous system.

2 At rest, the membrane is kept polarised at −60 mV inside the cell. When an action potential starts, it opens sodium ion channels and positively charged sodium ions diffuse into the cell. This raises the potential from −60 mV towards +40 mV, so the membrane is getting less polarised — this is depolarisation. When the potassium ion channels open, the positively charged potassium ions diffuse out of the cell. This reduces the +40 mV charge back to −60 mV — this is repolarisation.

3 When an action potential arrives at a node of Ranvier, it opens sodium ion channels. Sodium ions diffuse through the membrane into the cell. As they enter the cell, the concentration of sodium ions inside the cell increases and the area becomes more positive. This causes the ions to move along the neurone in the form of a local current. This local current carries the sodium ions along to the next node. When the sodium ions arrive at the next node, their charge causes voltage-gated sodium ion channels to open, starting the action potential at the next node.

3 Hormones

1 There may be many tissues affected by one hormone. For example, adrenaline affects the muscles, the heart, the blood vessels and the lungs. The hormone needs to be transmitted all over the body to reach all of these tissues. However, the hormone has a specific shape, which means it only binds to specific receptors with a complementary shape. Therefore, the hormone only binds to cells that have the correct membrane-bound receptor.

2 cAMP is the second messenger inside the cell. It activates enzymes inside the cell cytoplasm. The cell may be specialised to manufacture only certain enzymes. For example, in liver cells the cAMP activates phosphorylase enzymes that break down glycogen.

3 Stimulates the breakdown of glycogen — this releases glucose from glycogen and increases the concentration in the liver cells so that glucose enters the blood and is transported to the muscles.

Increases blood glucose concentration — the muscles have energy to contract.

Exam practice answers and quick quizzes at **www.therevisionbutton.co.uk/myrevisionnotes**

Increases heart rate — increases transport of oxygen and glucose to the muscles for increased aerobic respiration.

Increases blood flow to the muscles — to supply more oxygen and glucose for aerobic respiration.

Decreases blood flow to the gut and skin — so that more blood can flow to the muscles.

Increases width of bronchioles to ease breathing — so that more air can enter the alveoli to increase gaseous exchange.

Increases blood pressure — to speed up blood flow.

4

Feature	Beta cell	Synapse
Potential difference across membrane at rest	-60 to $-70\,mV$	-60 to $-70\,mV$
Potassium channels at rest	Potassium channels open	Potassium channels closed
Calcium channels at rest	Calcium channels closed	Calcium channels closed
Effect of stimulation on potential difference	Depolarisation to $-30\,mV$ by closure of potassium ion channels	Depolarisation to $+40\,mV$ by opening of sodium ion channels
Effect of depolarisation	Calcium channels open	Calcium channels open
Effect of calcium influx	Vesicles of insulin fuse to membrane	Vesicles of acetylcholine fuse to membrane

5 Someone with diabetes mellitus will be unable to convert glucose to glycogen. Therefore, there are no stores of glycogen in the liver or muscles. As a result, once glucose in the blood is used up there are no stores that can be converted to raise blood glucose levels. The muscles lack energy and cannot respire as much or make ATP for the muscles to use in contraction.

4 Excretion

1 Liver cells have no external specialisations. In their cell surface membrane they possess many channel proteins and receptor sites. The receptor sites are specific to adrenaline, insulin and glucagon — a different type of site for each hormone. Inside the cytoplasm are granules of glycogen to store glucose. There are many ribosomes to manufacture the wide variety of enzymes used in the cells.

2 Liver cells are very active and require oxygen for aerobic respiration, so oxygenated blood from the hepatic artery is important. The role of the liver is to treat the blood to remove impurities and toxins as well as excess substances that could be useful to the body. Therefore, it receives blood from the digestive system via the hepatic portal vein. This brings in blood carrying all the substances that have been absorbed from digestion.

3 Ammonia is highly toxic, so it cannot be allowed to build up in the tissues. It affects the action of enzymes by denaturing them. Mammals must conserve water as they only gain water through eating and drinking, so they cannot afford to dilute the ammonia a lot before excreting it. Therefore,

ammonia is converted to urea which is less toxic. Fish are surrounded by water and can simply dilute the ammonia so much that it is not very toxic and excrete it along with a lot of water.

4 Proteins are too large to pass out of the blood through the basement membrane, so they are not found in the tubule and do not need to be reabsorbed.

5 When the water potential rises above the set point, the mechanism has a reversing effect to bring the water potential back to the set point. If the water potential falls, the mechanism brings it back up to the set point. Any change away from the set point brings about a reversal of that change so that the water potential remains close to the set point.

5 Photosynthesis

1 White light consists of a range of wavelengths. Chlorophyll absorbs the red and blue wavelengths but reflects the green light — this is the light that enters our eye to enable vision.

2 The green chlorophyll breaks down before the accessory pigments. In autumn, the leaves change colour as the accessory pigments become visible — xanthophyll is yellow.

3 Phosphorylation is the name given to the addition of a phosphate group to ADP to produce ATP. If this happens using the energy from light, it is called photophosphorylation. After absorption the energy from light is held by excited electrons. These electrons pass their energy to the ATP molecule during phosphorylation. As the electron loses its energy, it passes from one carrier to another. If the electron ends back at its original place in a photosystem, it is known as a cyclic reaction (cyclic photophosphorylation). However, the electron may be passed to another molecule and not return to its original place, which is called a non-cyclic reaction (non-cyclic photophosphorylation).

4 Light is used to produce ATP and reduced NADP during the light-dependent stage. If the light intensity is low, less ATP and reduced NADP are produced. These compounds are used in the light-independent stage to convert GP to TP. If less GP can be converted (because there is less ATP and reduced NADP), GP will accumulate. In order to produce glucose, the GP must be converted to TP which is used to make glucose. Therefore, if less TP is produced, less glucose will be made.

5 (a) There must be enough oxygen in the water to prevent the oxygen released by the plant dissolving.

(b) Carbon dioxide in the water may be limited. Dissolving sodium hydrogen carbonate in the water ensures there is enough carbon dioxide to prevent it limiting the rate of photosynthesis.

(c) The time interval allows for the apparatus to equilibrate so that the rate of photosynthesis is constant for those conditions.

(d) The bubbles may be difficult to count accurately, especially if they are appearing quickly. Also the bubbles may not all be the same size.

6 Respiration

1 Hydrogen atoms are removed from the substrate to reduce NAD or FAD. The energy associated with that hydrogen is used to drive the electron transport chain, the pumping of hydrogen ions and subsequently the production of ATP. As there are more carbon–hydrogen bonds in fats than in carbohydrates, there are more hydrogen atoms available to reduce the NAD or FAD.

2 Coenzyme A must be released to pick up another acetate molecule and feed it into Krebs cycle.

3 Oxygen is the final electron acceptor. If electrons are not removed at the end of the electron transport chain, the chain will not transport more electrons. This means that no protons will be pumped across the inner mitochondrial membrane and no proton gradient is built up. Therefore, protons will not flow through ATP synthase and less ATP will be made.

4 The protons pumped across the inner mitochondrial membrane diffuse away and do not accumulate. Therefore, there is no proton gradient and no proton motive force to drive the ATP synthase activity.

5 Anaerobic respiration does not include Krebs cycle, oxidative phosphorylation or chemiosmosis. These processes are where most ATP is made.

7 Cellular control

1 There are 20 amino acids commonly used to build proteins. Therefore, there must be at least 20 codes. Reading a single base at a time gives 4 codes. Reading bases in pairs gives 16 codes. Reading triplets gives 64 possible codes. This is more than enough for the number of amino acids.

2 The DNA code is in the nucleus, but the ribosomes and the other requirements for protein synthesis are in the cytoplasm. The DNA molecule is too large to fit through the nuclear pores. mRNA is much smaller (because it is a copy of only part of the whole chromosome) and can fit through the nuclear pores.

3 The sequence of base triplets (codons) on mRNA carries the code for the sequence of amino acids in the protein to be made. Each triplet enables the binding of a specific tRNA molecule in the correct place so that the amino acids are aligned correctly. If the mRNA was double-stranded, the base sequence would be hidden.

4 If an organism is well adapted to its environment, it will be making proteins that are suitable to their role. Any mutation that causes a change to the structure of that protein is likely to make it less effective in that particular environment, so the mutation is harmful.

5 Energy is required to manufacture proteins. If there is no lactose in the environment, it is a waste of energy to manufacture β-galactosidase and lactose permease. The bacterium can conserve energy by not making these proteins.

8 Meiosis and variation

1 The DNA in the cell replicates once so that there is twice the normal amount of DNA. After the first division, which separates the homologous chromosomes, the cells are haploid but they still contain two copies of each chromosome. The second division separates the sister chromatids so that the cells have one copy of each chromosome. This means they have half the usual DNA content and are suitable for fertilisation, which will restore the diploid number.

2 (a) A dominant allele is one that is expressed in the phenotype and masks the expression of a second allele completely. A codominant allele is one that does not completely mask the second allele, so that both contribute to the final phenotype.

(b) A chromosome is a length of DNA wrapped around histone proteins. It contains a number of genes within its structure. After replication, a chromosome appears as two identical strands joined by a centromere and each strand is a chromatid.

(c) A gene is a length of DNA that carries the code for a polypeptide. An allele is an alternative version of a gene.

3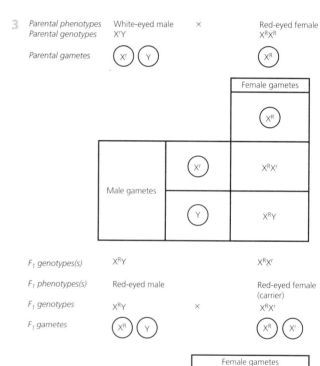

4
Parental phenotypes	Red Male	×	Roan female
Parental genotypes	RR		RW

Parental gametes

		Female gametes	
		R	W
Male gametes	R	RR	RW

F_1 genotype(s)	RR	RW
F_1 phenotype(s)	Red	Roan

No WW (White) individuals

5   $100 - 84 = 16\%$ show recessive characteristic. Therefore, $q^2 = 0.16$ and $q = \sqrt{0.16} = 0.4$.

$p + 0.4 = 1$. Therefore, $p = 0.6$.

The frequency of the heterozygous genotype is $2 \times 0.6 \times 0.4 = 0.48$. Therefore, 48% of the population are heterozygous.

9 Cloning in plants and animals

1   The conditions needed for the explants to grow include moisture, nutrients and warmth. These are ideal conditions for bacteria and fungi to grow, so if there are any microorganisms present they will grow quickly and destroy the explants.

2   Sexual reproduction involves meiosis to produce haploid gametes. Chiasmata during prophase introduce recombinations of the alleles. Also, fertilisation is random. Both these processes would change the genotype of the offspring and it would not be a clone.

10 Biotechnology

1   Initially conditions are ideal and there are no limits to growth and reproduction. However, as the population increases some factors required for growth and reproduction may become limited, such as oxygen, certain minerals or energy supply. This reduces the rate of growth of the population so the growth curve levels out. At the plateau the population is not growing any larger and the number of births equals the number of deaths.

2   Secondary metabolites are produced once the population is no longer growing. The population may be at a plateau or it may be declining. In continuous culture the population is maintained under ideal conditions that sustain exponential growth, so no secondary metabolites are produced. In batch culture the population is allowed to grow, plateau and decline. This when secondary metabolites will be produced.

3   The culture must be allowed to grow until the population is in the stable phase and then the conditions must be kept constant by adding just enough oxygen and nutrients and removing just enough of the culture and products to maintain the stable population.

4   Any trace of unwanted microorganisms must be removed so that only the required culture will grow. Unwanted microorganisms could compete for the nutrients, reducing the yield, or they could damage the product by metabolising it or contaminating it with toxic by-products.

11 Genomes and gene technologies

1   Each restriction endonuclease is specific to a certain sequence of nucleotides. If the same enzyme is used, it will cut at the same point each time. If different enzymes are used, they will cut in different parts of the DNA. When several identical long sections are cut up, this creates sections of different lengths that overlap. This allows overlapping regions to be used to piece together the whole sequence.

2   If two species are closely related to a common ancestor, they are not very different in evolutionary terms so their DNA will be similar. However, if two species are not closely related, it means that they have evolved separately for a long time, so there have been more opportunities for mutations and their DNA will be more different.

3   Taq polymerase works well at 72°C, which is unusually high for an enzyme. It has a tertiary structure that is stable even at high temperatures.

4   A sticky end is a sequence of exposed nucleotide bases. This allows a complementary sequence to bind in place. It ensures that the correct piece of DNA adheres to the plasmid or chromosome vector.

5   The section of DNA placed into a plasmid contains two genes for resistance to antibiotics. One gene is complete and confers resistance to the antibiotic. However, the second gene is not complete as it has been cut open and the gene for the required substance has been inserted into it. Therefore, the gene for resistance to the second antibiotic is not complete and will not confer resistance.

6   Somatic cell gene therapy affects only the cells treated. Germ line gene therapy treats the fertilised egg or zygote (or all the cells in a very early embryo) so that all the cells in the body possess the modification, including the cells that produce eggs and sperm. Therefore, the genetic modification is found in the sex cells and can be passed to the next generation.

12 Ecosystems

1   During succession, more and more species are able to survive in the modified conditions. Just before the climax community is reached, there are many climax community species living in the area but there are also some sub-climax species left. These are outcompeted and the diversity declines a little as it matures to the full climax community.

2   If decomposition is too slow, minerals and elements will not be returned to the soil quickly. This reduces the ability of plants to grow and the community will be less productive.

3   These plants are legumes, which have root nodules containing nitrogen-fixing bacteria. The bacteria absorb

nitrogen and convert it to a form that can be used by plants to produce amino acids and proteins. As farmers harvest their crops, they take away the nitrates that have been used for plant growth and it is important to replace those nitrates to maintain soil fertility.

13 Populations and sustainability

1 Scramble competition relies on all individuals finding what they can to eat or finding nest sites by chance. Genetic variation means that some individuals may be slightly better than others at competing. These individuals survive and breed more, passing on their alleles, so the frequency of those alleles increases. Competition by dominance hierarchy means that the strongest, most dominant individuals breed and pass on their alleles. If one individual becomes dominant and is the only one to breed, there is likely to be less variation between individuals in the next generation.

2 Preservation simply keeps the environment as it is. If the ecosystem has already declined and biodiversity is being lost, preservation will not prevent further loss. It takes active intervention to prevent further decline and loss of biodiversity.

3 Each year a small part of the woodland is cut to ground level. Over the next few years the area that has been cut grows back, providing a different stage of regeneration each year. This increases the range of habitats available and the biodiversity is increased. Since a different part of the woodland is cut each year, every stage of regeneration is maintained every year. This allows plants and animals to migrate around the woodland, living in the habitat that most suits them. The growth of wood from coppiced stumps is very rapid, so sustainable production can be achieved. The range of habitats created can also be used for a range of recreational purposes.

14 Plant responses

1 A rapid movement of leaves in response to touch may help to avoid predation by scaring an insect or surprising a larger herbivore. This is seen in the sensitive plant *Mimosa*. Other plants such as the Venus fly trap may move to trap food. A slow response to touch can lead to the stem of a climbing plant winding around a support such as a branch of a tree. This supports the climbing plant so that it can grow into the light.

2 Growth away from shade and into more intense light enables the plant to gain more energy for photosynthesis.

15 Animal responses

1 The two parts work in opposite ways: the sympathetic system prepares the body for activity by increasing heart rate, breathing rate and metabolic rate, along with other processes that are needed to be active, whereas the parasympathetic system conserves energy and allows the body to slow down.

2 A muscle can contract or get shorter on its own, but it is unable to stretch out again on its own. Therefore, when a muscle contracts it must be stretched out by the action of another muscle. Muscles are arranged in pairs that work against each other over a joint.

3 The I band is the length of a sarcomere that contains only actin filaments. As the myosin heads move to slide the actin past the myosin, there is greater overlap between the filaments so the length of the sarcomere that contains only actin gets shorter. Equally, the H zone is where there is only myosin and no actin. As the filaments slide past one another, the overlap increases which, in turn, reduces the length of the sarcomere where there are only myosin filaments.

4 The electrical stimulation of the muscle runs through the membranes of the muscle. This stimulation needs to pass in all directions over the heart to ensure coordinated contraction — the cross bridges help to carry the excitation in all directions. Also, cardiac muscle in the wall of the heart needs to squeeze the heart chamber in all directions so that the pressure inside the chamber increases. If all the fibres ran in one direction, the chamber would change shape rather than reduce in volume to put pressure on the blood.

5 The nervous system provides a very rapid response, which may be essential to escape danger. However, nervous stimulation may not last long as synapses can run out of neurotransmitters. The endocrine system takes a little longer to have its effect but it remains effective for longer, thus allowing a prolonged period of activity that may be needed for flight or fighting off an attack.

16 Animal behaviour

1 A maggot typically lives on decaying organic matter such as meat. If the maggot is in the light, it must be on the surface of its food and may be visible to predators. If it moves away from light, it is less likely to be seen. Also, as it moves away from light, it may be moving into its food where it can feed easily.

2 A dog does not normally salivate at the sound of a bell: salivation is a response to the sight or smell of food. There would be no benefit to salivating at the sound of a bell, so it is not an inherited form of behaviour. If the dog associates the sound of a bell with food because of training, it is learnt behaviour.

3 If gambling releases dopamine and the dopamine causes the release of adrenaline, the gambler feels a thrill. If more adrenaline is released, the person feels more of a thrill and is more likely to become addicted. Using a drug to block the dopamine receptor sites reduces the release of adrenaline, so the gambler feels less of a thrill and is less likely to become addicted.